HUMAN FACTORS METHODS FOR IMPROVING PERFORMANCE IN THE PROCESS INDUSTRIES

HUMAN FACTORS METHODS FOR IMPROVING PERFORMANCE IN THE PROCESS INDUSTRIES

Center for Chemical Process Safety

Dan Crowl, Editor

Authors: Dennis Attwood, Paul Baybutt, Chris Devlin, Walter Fluharty, Gareth Hughes, Dan Isaacson, Phil Joyner, Eugene Lee, Don Lorenzo, Lisa Morrison, Bob Ormsby, Graham Reeves

CENTER FOR
CHEMICAL PROCESS SAFETY

An **AIChE** Industry
Technology Alliance

WILEY-
INTERSCIENCE

A JOHN WILEY & SONS, INC., PUBLICATION

A Joint Publication of the Center for Chemical Process Safety of the American Institute of Chemical Engineers and John Wiley & Sons, Inc.

Published by John Wiley & Sons, Inc., Hoboken, New Jersey.
Published simultaneously in Canada.

For general information on our other products and services or for technical support, please contact our Customer Care Department within the United States at (800) 762-2974, outside the United States at (317) 572-3993 or fax (317) 572-4002.

Wiley also publishes its books in a variety of electronic formats. Some content that appears in print may not be available in electronic format. For information about Wiley products, visit our web site at www.wiley.com.

Library of Congress Cataloging-in-Publication Data:

Human factors methods for improving performance in the process industries / Center for Chemical Process Safety.
    p. cm.
"Wiley-Interscience."
Includes bibliographical references and index.
ISBN-13 978-0-470-11754-5
ISBN-10 0-470-11754-0
1. Chemical plants—Safety measures. 2. Chemical processes—Safety measures. 3. Human-machine systems. 4. Industrial productivity. I. American Institute of Chemical Engineers. Center for Chemical Process Safety.
TP155.5.H86 2007
660'.2804—dc22   2006028495

Printed in the United States of America.

10 9 8 7 6 5 4 3 2 1

Contents

HUMAN FACTORS TOOL KIT

Facilities and Equipment

■■■■ Preface

For over 40 years, the American Institute of Chemical Engineers (AICHE) has been involved with process safety and loss prevention in the chemical, petrochemical, hydrocarbon processing and related industries. AICHE publications are information resources for chemical engineers and other professionals to better understand the causes of process incidents and offer ways to prevent them. The Center for Chemical Process Safety (CCPS), a directorate of AICHE, was established in 1985 to develop and disseminate information for use in promoting the safe operation of chemical facilities and processes with the objective of preventing chemical process incidents. CCPS activities are supported by the funding and technical expertise of over 80 corporations. Several government agencies and nonprofit and academic institutions also participate in CCPS endeavors.

CCPS established a multifaceted program to address the need for process safety technology and management systems to reduce potential exposures from chemical process incidents to the public, the environment, and personnel and facilities. These programs are supported by the direction of its advisory and management boards.

Over the past several years, CCPS has extended its publication program to include a Concept series of books. These books are focused on more specific topics than the longer, more comprehensive *Guidelines* series and are intended to complement them—this book is a member of the Concept series. With the issuance of this title, CCPS has published over 80 books.

In 1989, CCPS published the landmark *Guidelines for the Technical Management of Chemical Process Safety*. This was followed in 1992 by *Plant Guidelines for Technical Management of Process Safety* which presented information on how to effectively implement, at a plant level, the various components of process safety management systems.

In 1994 CCPS produced two publications related to human factors. The first, *Human Factors in Process Safety Management,* provides guidance on how human factors considerations are applied to key elements of process safety management systems. The second book, *Guidelines for Preventing Human Error in Process Safety,* provides the underlying principles and theories of the science of human factors, as well as their application to process safety and to the technical management of process safety systems.

This book, *Human Factors Methods for Improving Performance in the Process Industries,* is the result of the desire of the CCPS sponsoring companies to provide guidance for plant engineering staff on specific, effective, practical techniques and tools that can be applied to improve process safety—it is not geared towards the specialist. The main focus of this book is on human factors issues relating to the avoidance of major process incidents; however, ergonomic considerations can contribute to and play a role in these incidents and are therefore also addressed.

This book provides an overview of how a human factors program can be established and implemented. The main content is contained in a series of summaries on particular human factors topics that a plant engineer can apply without reading the entire book. Each summary contains a short introduction to the topic, a description of the tools available, an example with graphics and flowcharts, as needed, and references for further reading. It is hoped that this book will take the specialist subject of human factors and present it in practical terms that any engineer can apply to ultimately improve process safety performance.

■■■■ Acknowledgments

This Guidelines book was written by members of the Center for Chemical Process Safety (CCPS) Human Factors subcommittee. Dr. Daniel Crowl (Michigan Technological University) was the technical editor and Bob Ormsby of CCPS provided staff support. The subcommittee was chaired by Phil Joyner (The Keil Centre; formerly of BP) and included (in alphabetical order): Dr. Dennis Attwood (RRS Engineering), Dr. Paul Baybutt (Primatech), Chris Devlin (Celanese), Tony Downes (FMC), Dr. Walter Fluharty (Bayer Health Care), Dan Horowitz (CSB), Gareth Hughes (formerly of Technica), Dan Isaacson (Lubrizol), Eugene Lee (US EPA), Don Lorenzo (ABS Consulting), Steve Marwitz (Formosa), Lisa Morrison (PPG), Mark Ploof (Air Products), Graham Reeves (BP), Dr. Henry Romero (PHS Concepts, Inc.), Fran Schultz (GE), Gary Staton (duPont), Stephen Werner (Intel), and Gary York (Rhodia).

Special acknowledgments are due to the primary authors of specific chapters in the book:

Dr. Dennis Attwood—Chapters 8, 15, 21, 22

Dr. Paul Baybutt—Chapters 4, 6, 7, 9, 10, 12, 14, 24, 25, 26, 27, 28

Chris Devlin—Section 2.4

Walter Fluharty—Chapter 20

Gareth Hughes—Sections 1.1, 1.2, 1.3, 2.1, 2.2, Chapter 23

Dan Isaacson—Sections 2.3, 2.4

Phil Joyner—Chapter 19

Eugene Lee—Introductions to the Human Factors Tool Kit

Don Lorenzo—Chapters 5, 13, 16, 31, Appendix A

Lisa Morrison—Sections 2.5, 3.2

Bob Ormsby—Sections 1.5, 3.1

Graham Reeves—Chapters 17, 18

Dr. Henry Romero—Chapters 11, 30

with assistance from Mrs. Elizabeth Romero

Gary York—Section 3.3

Important contributions to the writing of various chapters of the book were made by Tony Downes, Mark Ploof and Gary Staton.

CCPS also gratefully acknowledges the comments submitted by the following peer reviewers:

Greg Barrett—Lubrizol
Rick Curtis—The Dow Chemical Company
Pete Lodal—Eastman Chemical

Abbreviations and Acronyms

ABC	Antecedents, Behavior and Consequences
ADA	Americans with Disabilities Act
AICHE	American Institute of Chemical Engineers
API	American Petroleum Institute
BBS	Behavior Based Safety
CAS	Circadian Alertness Simulator
CCPS	Center for Chemical Process Safety
CCTV	Closed-Circuit Television
CFR	Code of Federal Regulations
CRO	Control Room Operator
DCS	Distributed Control System
DMQ	Dutch Musculoskeletal Questionnaire
E/E/PE	Electrical, Electronic and Programmable Electronic
EPA	U.S. Environmental Protection Agency
EPSC	European Process Safety Centre
FAID	Fatigue Audit Interdyne
FFD	Fitness for Duty
FI	Fatigue Index
FO	Field Operator
HAZID	Hazard Identification
HAZOP	Hazards & Operability Study
HCI	Human Computer Interface
HEART	Human Error Assessment and Reduction Technique
HF	Human Factors
HFAM	Human Factors Assessment Model
HFTD	Human Factors Tracking Database
HMI	Human/Machine Interface
HOF	Human and Organizational Factors
HRA	Human Reliability Analysis
HSE	Heath & Safety Executive (UK)
HVAC	Heating, Ventilation and Air Conditioning
IDEAS	Influence Diagram Evaluation and Assessment System
IEEE	Institute of Electrical and Electronics Engineers
IRIS	Incident Reporting Information System
ISO	International Organization for Standardization

KPI	Key Performance Indicator
KSA	Knowledge, Skills and Abilities
LOPA	Layer of Protection Analysis
MAC	Manual Handling Assessment Charts
MI	Mechanical Integrity
MOC	Management of Change
MORT	Management Oversight and Risk Tree
NASA TLX	NASA Task Load Index
NC	Noise Criterion
NPD	Norwegian Petroleum Directorate
OGP	International Association of Oil and Gas Producers
OSHA	U. S. Occupational Safety and Health Administration
OWL	Operator Workload
PHA	Process Hazards Analysis
P&ID	Piping and Instrumentation Diagram
PLC	Programmable Logic Controller
PM	Preventative Maintenance
PPE	Personal Protective Equipment
PROCRU	Procedure Oriented Crew Model
PSF	Performance Shaping Factor
PSM	Process Safety Management
PTW	Permit to Work
QEC	Quick Exposure Check
RAM	Reliability, Availability and Maintainability
REBA	Rapid Entire Body Assessment
RCM	Reliability Centered Maintenance
RMP	EPA Risk Management Program
RULA	Rapid Upper Limb Assessment
SAFE	System for Aircrew Fatigue Evaluation
SAFTE	Sleep Activity Fatigue and Task Effectiveness
SCADA	Supervisory Control and Data Acquisition
SCBA	Self-Contained Breathing Apparatus
SCMM	Safety Culture Maturity Model
SHE	Safety, Health and Environment
SLIM	Success Likelihood Index Methodology
SPL	Sound Power Level
SWAT	Subjective Workload Assessment Task
SWP	Safe Work Practice
SWORD	Subjective Workload Dominance
TAWL	Task Analysis/Workload
THERP	Technique for Human Error Rate Prediction
TLAP	Time-Line Analysis and Prediction
VDU/VDT	Visual/Video Display Unit/Terminal
W/INDEX	Workload Index
WRMSD	Work-Related Musculoskeletal Disorder
3DSSPP	3D Static Strength Prediction Tool

CHAPTER 1

Introduction

"... we must bear up against them (troubles), and make the best of mankind as they are, since we cannot have them as we wish."

— George Washington, Letter to General Philip Schuyler, Dec. 24, 1775

1.1 PURPOSE OF THIS BOOK

Consideration of the "human factor" is a key aspect of process safety management. Yet many organizations are struggling with the subject in order to improve process safety. The main purpose of this *Concept* series book is to assist organizations in their efforts to address human factors issues. Specifically, this book provides:

- A model and working definition of human factors;
- Illustrations of the benefits of human factors through the use of short case studies;
- Links between human factors and the business life cycle;
- Descriptions of key human factors topics together with practical techniques and tools that can be applied in the workplace at the level of both the corporation and the facility; and
- References for additional resources on human factors.

This book provides a basic understanding of the important elements of human factors and guidance for practical application. It will be useful to process and plant design engineers, plant operations and maintenance personnel, plant managers, and all other interested parties.

1.2 HUMAN FACTORS

Human factors is a common term given to the widely-recognized discipline of addressing interactions in the work environment between people, a facility, and its management systems.

Christensen, et al. (1988) defines human factors and ergonomics as:

"that branch of science and technology that includes what is known and theorized about human behavioral and biological characteristics that can be validly applied to the specification, design, evaluation, operation, and maintenance of products and systems to enhance safe effective, and satisfying use by individuals, groups, and organizations"

The UK Health and Safety Executive (HSG48, 1999) defines human factors as:

"environmental, organizational and job factors, and human and individual characteristics which influence behavior at work in a way which can affect health and safety."

The International Association of Oil and Gas Producers (OGP, 2005) defines human factors as:

"the interaction of individuals with each other, with facilities and equipment and with management systems."

A basic model for human factors relative to the process industries is shown as Figure 1-1. This model is based on the OGP model (OGP, 2005). This model identifies three domains for human factors: Facilities and Equipment, People, and Management Systems. These domains overlap each other and cannot be separated or removed from the model.

The Facilities and Equipment domain includes consideration of physical characteristics and work space, design and maintenance of equipment, and reliability.

The People domain includes consideration of individual attributes, skills, perceptions, and factors relating to fitness, stress, and fatigue. Some attributes, such as personality, cannot be changed, while other skills and attributes can. Computers and control systems play a major role in the safe and reliable operation of plants in the process industries. The interaction between humans (People domain) and computers (Facilities and Equipment domain) is one aspect of the overlap between domains.

The Management Systems domain provides the framework under which work is carried out. It includes procedures, training, process safety related work systems, and aspects of safety culture. Overriding all of these domains is the cultural and working environment. There are national, local, and workplace cultures as well as social and community factors.

Ergonomic issues are discussed repeatedly throughout this book. There are two major areas in ergonomics. The first area is ergonomics that results in personal injury or illness. The second area is ergonomics that results in human error. The second area is more the subject of this book than the first.

There are many aspects to human factors, but they all have the same goal—to fit the task and environment to the person rather than forcing the person to significantly adapt in order to perform the work.

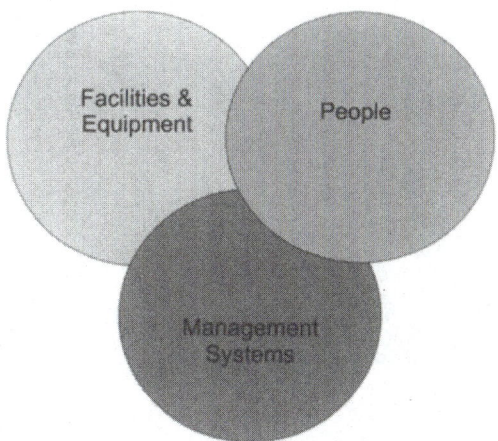

Facilities and Equipment Chapters

4   Process Equipment Design
5   Process Control Systems
6   Control Center Design
7   Remote Operations
8   Facilities and Workstation Design
9   Human Computer Interface
10 Safe Havens
11 Labeling

People Chapters

12 Training
13 Communications
14 Documentation Design and Use
15 Environmental Factors
16 Workloads and Staffing Levels
17 Shiftwork Issues
18 Manual Materials Handling

Management Systems Chapters

19 Safety Culture
20 Behavior Based Safety
21 Project Planning, Design and Execution
22 Procedures
23 Maintenance
24 Safe Work Practices and Permit to Work Systems
25 Management of Change
26 Qualitative Hazard Analysis
27 Quantitative Risk Assessment
28 Safety Systems
29 Competence Management
30 Emergency Preparedness and Response
31 Incident Investigation

Figure 1-1: A model for human factors based on the OGP model (OGP, 2005).

1.3   HUMAN ERROR

We often hear the term "human error" implicated as a cause of major accidents. In reality, all accidents can be attributed to human error due to errors in design, construction, operation, or maintenance. However, human error rarely refers to a single incorrect action by an operator or controller, as Reason's (1990) Swiss cheese model shows in Figure 1-2. Figure 1-2 shows that for every existing hazard there are several layers of protection (shown as Swiss cheese slices) that prevent that hazard from resulting in an accident. However, if the layers of protection fail (as shown by the holes), then an accident will result.

Humans have an influence over every layer of protection within a system. The actions of the last person at the incident end in Figure 1-2 may prove to be critical in

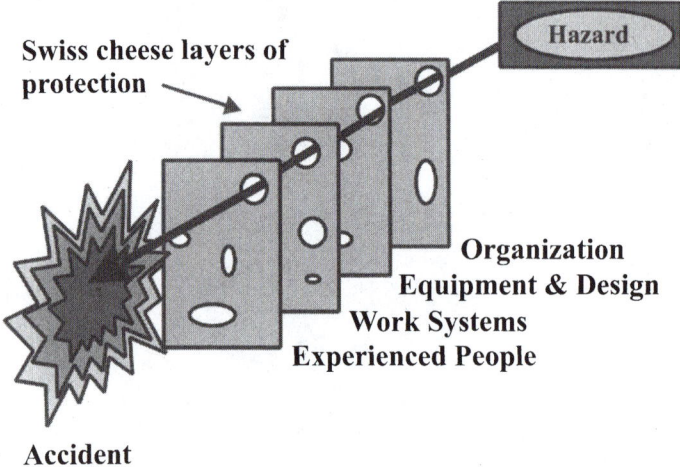

Figure 1-2: Reason's (1990) Swiss cheese model for the cause of accidents related to human error.

an accident sequence, but the same person has previously acted only as a gatekeeper in a more systematic process, preventing the previous errors or deficiencies of others from taking effect (e.g. poor design, poor decisions, lack of resources). Therefore, it is useful to understand where the holes/deficiencies lie and where simple human fallibility can prove critical to safe operations.

Types of Human Error

Reason (1990) outlined several types of human failure: slips, lapses, mistakes, and violations.

Slips are associated with faulty action execution, where actions do not proceed as planned. In this case the person knows exactly what they intend to do but it goes awry somewhere along the way. An example is an operator typing in the wrong number into a control system console, even though he intended to enter the correct number.

Lapses are associated with failures of memory. Again, the intention is correct but actions are omitted or parts of the action sequence are repeated. These errors tend to occur during the performance of fairly automatic or routine tasks in familiar surroundings, and attention is captured by something other than the task in hand. Examples include misreading a display, forgetting to press a switch, or accidentally batching the wrong amount to a batch reactor.

Mistakes occur when the execution is perfect but the plan itself failed to meet its objective, either because it was inappropriate for the situation, or the situation was novel and therefore no plan was available. An example is misdiagnosing the inter-action between several process variables and then carrying out incorrect actions.

Violations are situations where operators deliberately carry out actions that are contrary to organizational rules and procedures. This is not the same as sabotage—violators do not intend to cause harm. The operator might not follow the procedure because, for example, the procedures might be out-of-date or impractical. It is also possible that the procedure might appear to have little worth to the operators or the supervisors might condone doing things in a different way.

The above classifications for human error are useful since it allows us to predict the variations in human action within a task and to identify where the holes in the final slice of the Swiss cheese lie.

A temperature sensor failed on a large semi-batch polymer reactor. Unfortunately, it indicated a constant temperature which coincidentally was the desired set-point temperature—the reactor was actually at a much lower temperature. When the problem was eventually discovered, the operators decided to increase the temperature. Unfortunately, due to the low temperature the reaction rate was inadequate to consume the monomer feed and the resulting monomer concentration was very high. When the temperature was increased a reaction runaway occurred and the reactor exploded.

1.4 STRUCTURE OF THIS BOOK

This book is divided into two major parts. The first part is composed of Chapters 1 through 3. This part describes the purpose of the book, presents a model for understanding human factors, presents the need for human factors, and describes methods for implementing human factors in the corporate and facility setting. The second part presents a series of short, stand-alone summaries on particular human factors topics. Each summary contains a short introduction to the topic, a description of typical issues and examples, and approaches used to handle the topic. References of important topics and resources for more information are provided.

Chapter 3 addresses strategies for implementing human factors at the corporate and at the plant or engineering department levels. The emphasis is on identifying the required elements and the strategies for developing a human factors program. Section 3.1 and 3.2 deal with corporate level applications of human factors, stressing the need for management commitment, accountability, and clear measures of success. Section 3.3 and 3.4 presents a human factors implementation strategy for a plant or engineering department, including approaches for securing resources, establishing processes and continuous improvement.

The second part of this book introduces key topics on human factors identified by the CCPS Human Factors subcommittee. These topics are organized according to the major areas identified in Figure 1-1: facilities and equipment; people; and management systems. A broad range of human factors topics are covered in a concise and systematic format. Each of the topics is presented with an introduction, a focus on relevant human factors principles, available tools and industry applications, and key references. Where appropriate, graphics are used to reinforce concepts. References are provided for further information—including references that are cited directly in the text and additional references that provide more general background. Topics with potential regulatory implications (e.g., hazard analysis and OSHA PSM) are noted as such. Readers may go directly to their topics of interest as each major chapter in the second part is written to serve as a stand-alone description of the topic. The practical approaches to human factors presented in Part III will enable plant personnel to directly apply human factors topics of interest to improve process safety performance.

Table 1-1 links the topics in Part III to the OGP Human Factors model domains shown in Figure 1-1. Some of the topics obviously overlap with one or more domains.

The appendices contain a glossary of human factors terms and references for further information.

1.5   LINKAGE TO OTHER CCPS PUBLICATIONS

Human Factors has been addressed in many CCPS publications over the years because it is an important consideration in almost everything we do. CCPS sponsored the "International Symposium on Hazard Identification and Risk Analysis, Human Factors and Human Reliability" (CCPS, 1992a) and its proceedings were published by CCPS. Concurrently, the Human Reliability Subcommittee of CCPS developed and produced *Guidelines for Preventing Human Error in Process Safety* (CCPS, 1994a). This was a comprehensive text which provided basic concepts, principles and theories about human error, its application to process safety (including how it could be analyzed), and case histories. This was quickly followed by an addendum to the *Plant Guidelines for Technical Management of Chemical Process Safety* (CCPS, 1992b) entitled *Human Factors in Process Safety Management* (CCPS, 1994b).

The subject of human reliability analysis is very briefly addressed in the *Guidelines for Hazard Evaluation Procedures, 2nd Edition* (CCPS, 1992c). In *Guidelines for Chemical Process Quantitative Risk Analysis,* both the 1st and 2nd editions, (CCPS, 1989; 2000) a more detailed presentation is provided on various models and techniques for quantitatively analyzing human reliability.

Guidelines for Investigating Chemical Process Incidents, both the 1st and 2nd editions, (CCPS, 1992d; 2003) deals directly with human performance and human factors. Although individuals make mistakes it is usually because human factors issues were not properly addressed. This book recognizes those deficiencies and shows how to uncover them and to reveal the real root causes related to human factors.

Table 1-1: Linkage to the OGP domains, showing the overlap between the domains

Chapter	Facilities and Equipment	People	Management Systems
4. Process Equipment Design	✓	✓	
5. Process Control Systems	✓		
6. Control Center Design	✓	✓	
7. Remote Operations	✓	✓	
8. Facilities and Workstation Design	✓	✓	
9. Human / Computer Interface	✓	✓	
10. Safe Havens	✓	✓	
11. Labeling	✓	✓	
12. Training		✓	✓
13. Communications		✓	✓
14. Documentation Design and Use	✓	✓	
15. Environmental Factors		✓	
16. Workloads and Staffing Levels		✓	
17. Shiftwork Issues		✓	
18. Manual Materials Handling		✓	
19. Safety Culture		✓	✓
20. Behavior Based Safety		✓	✓
21. Project Planning, Design and Execution		✓	✓
22. Procedures			✓
23. Maintenance		✓	✓
24. Safe Work Practices and Permit to Work Systems			✓
25. Management of Change			✓
26. Qualitative Hazard Analysis			✓
27. Quantitative Risk Assessment			✓
28. Safety Systems	✓		✓
29. Competence Management		✓	✓
30. Emergency Preparedness and Response	✓	✓	✓
31. Incident Investigation			✓

1.6 REFERENCES

CCPS (1989), *Guidelines for Chemical Process Quantitative Risk Analysis* (NY: AICHE Center for Chemical Process Safety).

CCPS (1992a), International Symposium on Hazard Identification and Risk Analysis, Human Factors and Human Reliability in Process Safety (NY: AICHE Center for Chemical Process Safety).

CCPS (1992b), *Plant Guidelines for Technical Management of Chemical Process Safety* (NY: AICHE Center for Chemical Process Safety).

CCPS (1992c), *Guidelines for Hazard Evaluation Procedures, 2nd ed.* (NY: AICHE Center for Chemical Process Safety).

CCPS (1992d), *Guidelines for Investigating Chemical Process Incidents* (NY: AICHE Center for Chemical Process Safety).

CCPS (1994a), *Guidelines for Preventing Human Error in Process Safety* (NY: AICHE Center for Chemical Process Safety).

CCPS (1994b), *Human Factors in Process Safety Management.* (NY: AICHE Center for Chemical Process Safety).

CCPS (2000), *Guidelines for Chemical Process Quantitative Risk Analysis, 2nd ed.* (NY: AICHE Center for Chemical Process Safety).

CCPS (2003), *Guidelines for Investigating Chemical Process Incidents, 2nd ed.* (NY: AICHE Center for Chemical Process Safety).

Christiansen, J. M., Topmiller, D. A., Gill, R. T. (1988), "Human Factors Definitions Revisited," *Human Factors Society Bulletin,* v. 31, pp. 7–8.

HSG 84 (1999) *Reducing Error and Influencing Behaviour,* (Sudbury, UK: HSE Books), ISBN 0 7176 2452 8

OGP (2005) *Human Factors* (London, UK: International Association of Oil and Gas Producers, available at info.ogp.org.uk/hf/.

Reason, J. (1990). *Human Error.* (Cambridge: Cambridge University Press).

Reason, J. (1997) *Managing the Risk of Organizational Accidents* (Aldershot, UK: Ashgate Publishing).

CHAPTER 2

The Case for Human Factors

2.1 WHY IS HUMAN FACTORS NEEDED?

Human factors plays an important role in process safety by utilizing scientific knowledge and fundamental principles from many disciplines to reduce the frequency of accidents. Its proper application can also result in improved operating effectiveness and reliability.

With the application of human factors it is possible to reduce the likelihood of human error, increase productivity and quality, and reduce the risk of work-related musculoskeletal disorders. If human factors principles are proactively integrated throughout the lifecycle of an asset—from the design stage, through commissioning, operation and maintenance and looking ahead to decommissioning—then the people within the system will respond in a proactive manner rather than simply reacting to problems that arise.

Quite simply, human factors must be considered because the success of any organization depends on the performance of humans. Virtually every facet of any industrial organization involves human factors because the organization will only survive if humans successfully produce a product or deliver a service. At its most basic level, the organization must consider human factors as it hires employees—to ensure they have the required physical and mental capabilities to perform the required job tasks. Then the organization must consider human factors as it designs and/or procures the physical assets its employees will use to perform the required tasks—to ensure the tools, procedures, and facilities are conducive to successful human performance. Finally, the organization must consider human factors in its own organizational structure—to ensure that the management structure and the relationships between work groups are conducive to successful human performance.

Unfortunately, the need to consider human factors is often only recognized after a spectacular failure occurs, such as those described in Chapter 3. Rather than waiting to appear in the headlines, proactive companies aggressively use their incident investigation system to identify near miss situations in which human errors occurred but did not result in serious consequences this time. They use those near misses to identify error-likely situations and correct the underlying human factors deficiencies, which may be as simple as a missing label or as complex as a breakdown in communications between work groups.

It is surprising that the value of human factors has not yet been comprehensively accepted throughout the industry. Some misperceptions for this may be:

- Human factors is ill-defined;
- Human factors is difficult to apply and therefore is only the domain of specialists;
- Human factors is expensive and adds little tangible benefits

Each of these issues is discussed below.

Human Factors is Ill-Defined

Many definitions are available for human factors, several of which are provided in Chapter 1, Section 1.2, so human factors is actually well-defined. The problem is actually one of application and practice. For most technical areas, like process control, it is clear where the technical area must be applied. But for human factors this is not so clear. The essential question is: When should human factors issues be considered? In reality, human factors issues are important in all areas where humans are involved. The broad domain of human factors contributes to the misconception that it is ill-defined.

Human Factors is Difficult to Apply

The broad domain of human factors also contributes to the mis-conception that it is difficult to apply. Since the domain of application is so broad, a wide-range of methods have been developed. The large number of methods contributes to the mis-conception.

Furthermore, human factors involves a number of "soft skills" such as safety culture and training. For engineers trained in more fundamentally based methods, these may appear difficult to apply. This also contributes to the mis-conception.

Human factors, like most disciplines, has a wide range of problems that vary in both scope and technical difficulty. Some of these problems may be solved through the use of relatively simple techniques, which, following some training and practice, can be readily applied by non-specialists.

Human Factors is Expensive

Section 2.2 contains a number of case histories related to human factors. Many of these incidents resulted in human injury, death, damage to the environment, loss of capital equipment and inventory and damage to the industry's license to operate. On a daily basis, human factors issues cause ruined batches, off-spec products, and unplanned shutdowns. Clearly the costs of these accidents far exceed the proper application of human factors in the first place.

More detailed case histories showing the benefits to human factors is provided in Section 2.3: Business Value/Justification.

So why is human factors needed? Because proper attention and investment in human factors pays enormous dividends in lives saved, environmental incidents avoided, and bottom-line gains in profit for the enterprise.

A case study that incorporated human factors into their health, safety, and environmental processes showed that the combined employee and contractor recordable incident rates were measurably reduced (ExxonMobil, 2000) over a three year period. This improvement alone resulted in savings through reduced medical expenses, lower workers compensation and insurance costs, fewer replacement workers, and less equipment downtime.

2.2 PAST INCIDENTS

Human factors have contributed to a number of major industrial accidents with many leading to multiple fatalities.

Table 2-1 is a sampling of major industrial accidents that have occurred in the past 30 years or so. Table 2-1 identifies the accident and its consequence, and also lists the human factors elements that contributed to the accident. These include: communications, training, equipment, maintenance, decisions, procedures, and human machine interface.

2.3 BUSINESS VALUE/JUSTIFICATION

There is no question that business demands require us to reduce incidents and environmental impacts and still increase productivity. We have seen success in reducing incident frequency by adopting improved engineering solutions and management systems. How can we achieve further improvements in process safety performance? One solution is shown in Figure 2-1. To accomplish additional improvements, human factors must be effectively applied and managed within your organization. This will require resources and management commitment but will result in significant performance improvements.

What are the benefits of taking human factors into account within your organization?

- Fewer accidents
- Fewer near misses
- Reduced potential for human error and its consequences
- Improved efficiency
- Lower lifetime costs associated with maintaining and re-engineering systems
- A more productive workforce

A few case histories will demonstrate these benefits.

BP, at its Grangemouth facilities (Joyner, 2004) identified an overwhelming majority of incidents attributable to human error. Several initiatives were launched to improve human factors at the individual, task, and organizational levels. Over a 3-

Table 2-1: Summary of significant accidents with human factors issues

Accident	Human Consequence				Human Factors Issues				
		Communications	Training	Equipment	Maintenance	Decisions	Procedures	Human Machine Interface
Flixborough, England, 1974: Cyclohexane release. This accident resulted in an inquiry and demonstrated the criticality of good management of the change process	28 fatalities 86 injured	Drawing of change done on shoproom floor only.	The operators were not trained in hazard identification.		Not supervised by qualified engineer.	To bypass one reactor vessel with a temporary modification to allow the plant to continue to operate.	Poor management of change. No time limit set for the temporary change.	
Seveso, Italy, 1976: Dioxin release. This accident resulted in creation of the European "Seveso Directives."	No fatalities directly attributed. Multiple illnesses	Company did not communicate which chemicals had been released.		Bursting disc blew on a reactor vessel —set point too high. Reactor vessels inadequate.		Secondary receiver recommended by manufacturer to collect any vented material—not fitted.	Failure to follow operating procedures—Batch not finished and operation not shut down per normal shutdown procedures because of weekend holiday	
Three Mile Island, Harrisburg, PA, 1979: Loss of control of a nuclear reaction resulting in destruction of the reactor core. This accident halted expansion of the nuclear industry in the U.S.	None	A near miss at another unit was not communicated to this unit.	Training for operators not adequate—no feedback to students.	Turbine trip. Subsequently, PORV sticks open.	Two block valves left in closed position after maintenance 2 days before.	Operators reduced coolant water flow into reactor attempting to prevent flooding—caused meltdown.		Operators misled by control panel—poor design. Over 100 alarms—not prioritized. Warning light showing valves closed obscured by maintenance tag.

(continued)

Accident	Fatalities							
Newfoundland, Canada, 1982: Ocean Ranger Platform collapse	84 fatalities	Ocean Ranger's radio operator did not call for help on normal distress frequencies. Instead, he used a frequency used exclusively by oil companies.	Only 40 of the 200 crew who worked the rig had been trained in the use of lifeboats.	"Useless safety gear"—no survival suits available. Lifeboats not designed for waves that high.			Control room operators had little training and no procedures.	
Bhopal, India, 1984: Release of Methyl Isocyanate (MIC). This accident was the worst single incident in history of the chemical industry.	Est. 8,000 fatalities 300,000 injured	No alarm ever properly sounded to warn of gas cloud. Failure to provide MIC treatment information.	Half to two-thirds of skilled engineers had left prior to the accident.	Insufficient scrubber capacity. Flare tower disconnected. Vent gas scrubber in inactive mode. No gas masks available.	Pressure and temperature sensors did not work—pressure gauge under-reading by 30psig. Refrigeration plant shut down to save costs. No regular cleaning of pipes and valves.	To store 10 times more methyl isocyanate than required on site.	Poor evacuation measures. No temporary management of change.	No online monitor for MIC tasks. No automatic sensors to warm of temperature increase.
Pasadena, TX, 1989: Flammable vapor release and explosion. This accident caused OSHA to consider contractor safety more heavily in their PSM Standard.	23 fatalities 132 injured	Prior concerns that alarm was not audible in certain areas of the plant. Emergency control room damaged by blast so communications disrupted.		Part of plug lodged in pipework during clearing work.	Specialist contractor employed but no site personnel in attendance throughout. Air hoses to valve cross-connected so opened when should have closed.	Only single isolation provided for this operation.	Isolation and double block procedure not followed on site. Procedure required air hoses to be disconnected prior to maintenance—not done. Permit to work system not enforced.	

Table 2-1: Summary of significant accidents with human factors issues *(continued)*

Accident	Human Consequence	Human Factors Issues						
		Communications	Training	Equipment	Maintenance	Decisions	Procedures	Human Machine Interface
Milford Haven, UK, 1994: Major fire	26 injuries	Alarm flooding—one every 2 to 3 secs. Senior personnel joined operators in CCR—too many cooks not enough coordination.	Poor training relating to abnormal operations.	Flare drum pump-out system modified three years earlier—HAZID identified problem but no further analysis performed.	Poor maintenance regime—many faults were being "lived with" until planned shutdown.	To keep unit running when it should have been shut down.	Shutdown of neighboring units according to procedure—successful.	Control valve indicated open when it was in fact shut. Too much text, not enough use of color on screens. No overview of whole process.

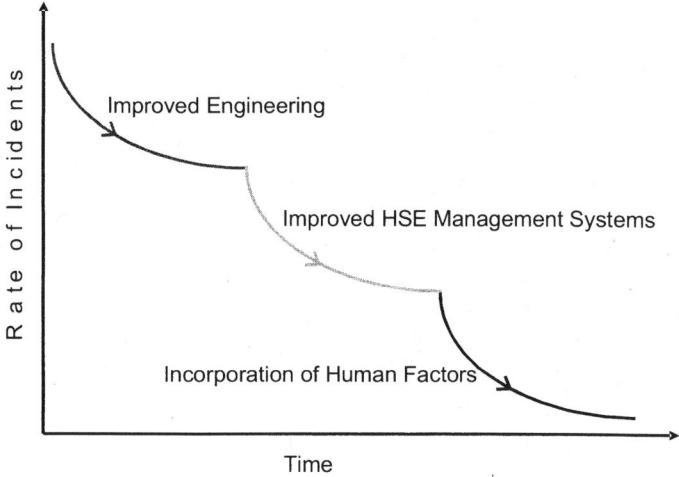

Figure 2-1: A plot showing how human factors can reduce process safety incidents. Human factors can improve process safety after improved engineering and management systems have achieved their results.

year period, as a result of these initiatives, there was a 10% improvement in plant reliability (from 85% to 95%) and a 25% reduction in costs (off-specification products, rework, etc.). Even more importantly, the plant's safety metrics showed a 200% improvement for the period.

Air Products and Chemicals, Inc. (Ploof, 2003) had similar success in reducing the number of controllable outages in its large hydrogen/carbon monoxide units worldwide. A human factors improvement program was implemented after incident investigations showed an unacceptable upward trend in accidental plant trips. Over a 2-year period, the program yielded a 60% reduction in the number of controllable outages. This increased revenue by millions of dollars versus the few hundred thousand invested in the improvements.

ExxonMobil Inc. (ExxonMobil, 2000) was also successful in implementing human factors improvements for its plant maintenance group at its Baton Rouge chemical facilities. Numerous improvements were made to facilities, tools, and work procedures to minimize manual handling, climbing, and overexertion, and to ensure that necessary information was readily available. Portable test stands were installed in all 72 substations to eliminate the need to test breakers on line, reducing risks of electrical injuries and outages. Electric lifting devices installed in trucks and switchgear eliminated manual handling of breakers that could lead to back injuries. Proper power and hand tools reduced risks of overexertion. There were no recordable injuries in that work group in the five years after the effort began and unplanned outages were reduced by 80%.

Addressing human factors appropriately within your business/organization is a cost effective process safety element. Human factors apply to all areas of an organization and at all stages within the life of the business. Many project management teams have learned that the most cost effective opportunity to incorporate human factors into equipment and systems comes when a facility is initially designed. When human factors are addressed proactively during plant and work place design, costly rework is reduced, startups are smoother and incident frequency is lower.

Six important components of human factors implementation are common to organizations that have applied the technology successfully:

1. The efforts are driven by specific needs and opportunities for performance improvements.
2. Improvements are sustained by building them into existing management systems, standards, and practices.
3. Effective human factors resources and tools promote workforce awareness and aid implementation.
4. Roles and responsibilities for the efforts are clearly defined.
5. Results are evaluated and shared to enhance benefits and effectiveness.
6. Management is committed throughout the organization to continual human factors improvement.

Management leadership/commitment is key to successful human factors implementation. Management must be actively involved and needs appropriate training to identify and apply human factors within operations and throughout the organization. The organization cannot expect excellent performance by employees without addressing human factors issues. Including human factor elements in risk management is a valuable investment in overall process safety.

2.4 HUMAN FACTORS IN THE LIFE CYCLE OF A CHEMICAL PLANT

The typical life cycle of a chemical plant includes four broad categories:

1. Process Design and Project Execution
2. Plant Operation
3. Plant Maintenance
4. Decommissioning

It is most cost effective to address potential human factors issues in the initial design and construction of a chemical plant. Human factors issues addressed later in the life cycle will result in considerably more time, effort, and cost to handle. The cost will be particularly high if the issue is identified as a result of an accident. For example, if a project engineer provides too many alarms in a new control system, then the operator will be easily overloaded with alarms and may not be able to diagnose the signigicant fault properly. The cost of an accident will certainly exceed the cost of a proactive effort to sort out the unnecessary alarms.

It makes sense to continuously evaluate human factors throughout the operation and maintenance of a chemical processing facility to remove those human factor concerns in an on-going proactive manner. Human factors can be addressed proactively during the conduct of any of the myriad of safety reviews performed during plant operation, such as PHAs, MOC Safety Reviews, Job Safety Analysis, etc. Human factors issues can also be addressed reactively after a process safety incident—surely, the proactive approach is the best approach.

The Department of Defense (DOD) has designed a number of plants to destroy stockpiles of chemical weapons. DOD has actively integrated human factors into all phases of the plant operation including the ultimate demolition and removal of the plant materials and equipment.

Part III of this book discusses human factors using a *topical* organization. To assist the reader in understanding the life cycle approach, the plant life cycle organization is provided in Table 2-2, which clearly shows that human factors issues are important in all of the major life cycles of a chemical plant.

2.5 NEEDS DRIVEN PROGRAM

To ensure a successful roll-out of a human factors program one must consider the different needs of the various parts of the organization.

The first issue is that efforts toward implementing human factors must be tailored to the needs of each specific entity. That entity may be one specific site, an entire business unit, or the entire company, depending on how homogeneous the needs are. Different plant sites will have different needs based on the function of the site. The human factors requirements at a refinery are likely very different from a small batch chemical plant. Likewise, the human factors requirements at corporate headquarters, research and development, and production sites are likely to be different. In addition, both the geography and the country change the needs for a specific application of human factors. A facility in India will have different needs than one in Alaska, due to the differences in physical environment, culture, and regulatory aspects.

All of these differences lead to approaching human factors implementation from a needs-driven strategy, rather than a cookie-cutter one. The strategy should be bottom-up, focusing on those human factors activities yielding the highest benefits at the lowest costs, all with strong support from upper management.

To determine these needs, data should first be collected from a variety of information sources. Sources include worker feedback, behavior based safety observations, incident investigation reports (including near misses), injury and illness reports, absentee reports, product quality deviations, reliability data, and other reports of non-conformances. Every company has many sources of information, and these information sources can be diverse across different parts of the organization.

TABLE 2-2: Human factors in the life cycle of a chemical plant

Chapter	Design	Operation	Maintenance	Decommissioning
4. Process Equipment Design	✓	✓	✓	
5. Process Control Systems	✓	✓		
6. Control Center Design	✓	✓		
7. Remote Operations	✓	✓		
8. Facilities and Workstation Design	✓	✓		
9. Human/Computer Interface	✓	✓		
10. Safe Havens	✓	✓		
11. Labeling	✓	✓	✓	✓
12. Training		✓	✓	
13. Communications	✓	✓	✓	✓
14. Documentation Design and Use	✓	✓	✓	✓
15. Environmental Factors	✓	✓	✓	
16. Workloads and Staffing Levels	✓	✓	✓	✓
17. Shiftwork Issues		✓	✓	
18. Manual Materials Handling	✓	✓	✓	✓
19. Safety Culture		✓	✓	
20. Behavior Based Safety		✓	✓	
21. Project Planning, Design and Execution	✓			✓
22. Procedures	✓	✓	✓	✓
23. Maintenance	✓	✓	✓	
24. Safe Work Practices and Permit to Work Systems	✓	✓	✓	
25. Management of Change	✓	✓	✓	✓
26. Qualitative Hazard Analysis	✓	✓		
27. Quantitative Risk Assessment	✓	✓		
28. Safety Systems	✓	✓	✓	✓
29. Competence Management	✓	✓		✓
30. Emergency Preparedness and Response	✓	✓		
31. Incident Investigation		✓	✓	

One company built a new chemical complex in southeast Asia, comprised of several different units. The units were built by separate teams with separate engineering firms. One team required that all fire extinguishers be red (per U.S. convention) while another team had consulted local experts and had specified blue which was the *local* convention for extinguishers. Unfortunately, neither team had checked the extinguisher color at an existing, operating plant at the site. Even worse, the fire hose connections were different in the two units and some had to be changed! Bunker gear was ordered in typical American sizes and was far too big for most of the local response team-members. Human factors must address the specific needs of each local complex.

The data can then be analyzed to identify the human factors issues that are causing the most problems at a site. Tools such as human factors checklists, human reliability analysis, and other tools described elsewhere in this book can be used to support a needs-driven approach for implementation.

2.6 REFERENCES

ExxonMobil (2000) "Human Factors: A Guide to Action," (Irving, TX: Exxon-Mobil).

Joyner, P. (2004), Private communication, June 16.

Ploof, M. (2003), Private communication, December 19.

CHAPTER 3

Performance Measurement and Improvements

3.1 BUILDING IMPROVEMENTS INTO EXISTING SYSTEMS

Today, most corporations have extensive standards and procedures that establish design, operation, and maintenance requirements. Associated with the execution of these standards and procedures are work practices and management systems that ensure that all activities are completed successfully. In many cases the structure and details of the work practices and management systems are actually built into the requirements of the standards and procedures.

The easiest way to incorporate human factors within a company or organization is to build them into the existing standards and procedures, as opposed to creating new standards. This book provides ideas and examples where improvements in human factors can be realized. In fact, many operating or design companies already have standards and procedures associated with most of the topics presented in the book, and modifications can be made in most cases.

What exactly are the triggers causing human factors considerations and requirements to be incorporated into existing systems? A starting list includes: accidents and near miss investigation results, regulations, a proactive senior management, and employee involvement programs.

Accidents and Near Miss Incidents

Almost every accident or incident has failures in management systems that contributed to the event. Accident analysis usually uncovers many human factors deficiencies. The root cause of these deficiencies is typically inadequate application of human factors in the development or application of various standards, procedures and management systems. Or, the failure may result from improper execution of these standards, procedures, and management systems. If a company can identify gaps in these systems and correct them from a human factors standpoint, then the incident will be avoided the next time and many potential related incidents prevented as well. If the investigation and analysis is done well, it can be expected that many human factors issues will be identified in design, operation, and maintenance, all from this same incident.

Regulations

Federal and local governments are increasingly requiring consideration of human factors in regulations. The most obvious example is the OSHA PSM requirement which states that human factors must be incorporated into hazard reviews (OSHA PSM, 1992). The easiest approach to achieve these requirements is to modify existing standards and procedures that govern hazard reviews. By incorporating human factors into the hazard review process, the review team will have a much better understanding of possible causes (e.g. operator error) and assumed safeguards (e.g. operator response). This will increase the probability that the review will correctly judge the risks of hazardous events.

Proactive Management

The term "human factors" is becoming more widespread. As levels of senior management understand and become more familiar with this subject many will begin to appreciate the critical importance of managing this well. They will begin to establish and create the necessary resources and tools to leverage the associated benefits.

Employee Involvement

Employees are an excellent source for the identification of human factors issues and concerns, if given the proper atmosphere and encouragement to alert management. Suggestion systems, post-event analysis of activities like turnarounds, and other employee-based activities can uncover many potential improvements, which can be incorporated into management systems, operation directions, procedures, and so on to reduce the probability and/or consequences of a "human error."

> One company developed an extensive near miss incident investigation system that required the person submitting the report to check off from an extensive checklist all the human factor deficiencies that contributed to the incident. This information was automatically entered into a searchable database system. Each month the safety department queried the database and the results were used to identify opportunities for improvement.

3.2 MEASURES OF PERFORMANCE

In order to know if the organization's human factors program is effective, it is important to develop and use measures of performance. A time worn saying is "What gets measured gets done."

Most organizations have corporate goals and key performance indicators (KPIs) for safety incidents (recordables and lost time injuries), environmental incidents, near misses, plant shutdowns, and on-stream availability to name a few. All of these KPIs have direct links to human factors. Unfortunately, these are all lagging indicators and by the time they are noticed the harm/loss may have already occurred.

An organization's human factors program should have its own objectives, which must also be measured. If the program is done well, the results should lead to improvements in the organization's KPIs.

One objective might be to lower the frequency of ergonomic injuries through a specific company-wide ergonomics program. Possible measurements could include the number of people trained in the program, the number of potential injury-causing situations discovered and the number of corrective actions taken. Overall success can be measured by a reduction in ergonomic injuries, which ultimately leads to a reduction in the organization's recordable and/or lost time injuries.

A second objective might be to reduce the number of operational incidents or near misses involving the incorrect application (or non-application) of operating procedures. Possible measurements include categorizing and trending the basic causes for these failures before and after corrective measures are implemented.

It is important that corporate level decisions regarding human factors program coordination, resourcing, budgeting, and measurements defining success are fully developed and supported. Resources could include both corporate and external human factors expertise. Education of the entire organization about the program and its objectives is one of the keys to success. Everyone also must understand the goals and objectives of the program and the measurements used.

In order to track performance of the program, the measurement data should be easily captured and reported. The quality of the data obtained is directly related to training, familiarization of the program and the continued support of management. Ultimately, the KPIs for the organization should be positively affected.

A tool has been devised by the UK Health and Safety Executive (HSE, 2004) to measure the current state of a human factors program in an organization. This tool utilizes a Human Factors Assessment Model (HFAM). The aim of a HFAM assessment is to determine which of the following five levels of human factors capability best describes the organization being assessed:

- Not following good practice.
- Some elements of good practice achieved, but not enough to be confident that it will be applied consistently.
- Good practice.
- Good practice achieved, towards best practice.
- Best practice.

Although originally developed for the off-shore industry, this model has been shown to be applicable to any industry, that uses human factors to improve safety and reliability.

Another tool which might find application for specific components of human factors are a set of static and dynamic checklists which are included in Appendix A.

One company was experiencing a number of process incidents. They began to measure and track these incidents and noted that a large number of incidents were related to mechanical integrity (MI) programs and systems. They focused attention on their MI systems and saw measurable improvements.

3.3 ROLES AND RESPONSIBILITIES

Management commitment to human factors principles must be visible, steadfast, and obvious within the implementing organization. A successful human factors program demands that clear accountability for actions are embedded at all levels in the organization. The establishment of the organizational roles and responsibilities is necessary to drive that accountability as well as adherence to the standards required within the program.

Line management has the ultimate ownership, and thus responsibility, for any human factors effort. Each level in the line organization must clearly establish the program responsibilities for that effort. Line management is responsible for ensuring that the management system is developed, communicated, implemented, functioning, measured, reviewed, and improved.

Generally, line management will need support from human factors experts to implement a program. The experts may be internal or external. The experts will be responsible for specific training in the use of human factors tools, consultation, and advice on their use, guarantors of the method application, and, if requested, interpretation of results and recommendation development. An organization may choose to develop internal expertise at a given site or at a shared level. If so, adequate education must be provided to develop the necessary internal expertise. Alternatively, the organization may use external experts for assistance, in which case a liaison or coordinator for the expert's services is required.

Organization personnel, both hourly and salaried, also have key roles and responsibilities, which bear on program success. They are key contributors to the effort. They must actively participate in the various human factors activities. Their input during the human factors analysis is essential to understanding and defining the issues within the organization. They also have a prime role in developing solutions since they will ultimately implement and use the solutions.

A possible model for implementing human factors programs will consider how the program should accomplish the steps shown below. Clear responsibility for each step must be established (ExxonMobil 2000):

- Familiarize management with human factors considerations and gain their alignment on objectives.
- Identify and prioritize systems, practices, and procedures that would benefit most from improvement.
- Develop a plan and resources for action.

- Address specific needs and opportunities for performance improvement.
- Ensure that personnel directly involved in the implementation get appropriate training and support.
- Execute action plans, including the integration of human factors considerations into management systems, standards, and practices.
- Measure results.
- Share results, successes, failures, and technology recommendations.

One company asked its operators at each plant to complete a human factors survey. When all the results were tabulated the plant, through a special task group formulated its HF goals for the next year with both hourly and salary responsibilities and accountabilities defined.

3.4 CONTINUOUS IMPROVEMENT

The culture and leadership of an organization helps drive continuous improvement. How an organization applies human factors is closely related to their success.

Continuous improvement in any human factors program or activity, utilizing human factors principles, requires the leverage of lessons learned. How best to accomplish this varies within an organization, and sharing of the results is obviously an important part of the process. Sharing of successes within any company today is an easy method to promote human factors improvements. Typical examples of successes include new production records, time without a lost workday case, percentage of defect-free products, human factors program success stories, and so on. These successes can add to the pride employees take in their job, business, and company and can encourage people in an organization to "keep the record going." In addition, sometimes the successes of one part of a company can be used as motivation for the whole company. If one part of the organization can achieve why can't the other parts? How can we "bootstrap" the others to achieve the same success? The only caveat is that management must ensure that the success was not achieved by cutting corners or putting people at risk. The successes should not be obtained at the expense of safety.

Sharing of negative results with respect to human factors is much more difficult. Incident investigations typically fall under this category. Incidents are undesirable events and most people see these as failures. More enlightened companies use the opportunity to share the results and use the incident report as a positive for the benefit of other parts of the company. Fortunately, serious incidents are rare in a well-managed organization.

Near miss incidents are an excellent source of information about human factors. Near misses are less negative and, if analyzed constructively, can provide extremely valuable information and insights into human factors issues. This information must be shared effectively to achieve the maximum benefit. Thus, companies should include provisions for information sharing in their management system and ensure that changes are implemented, where applicable.

Incident and near miss investigations must be very thorough and must go beyond basic and contributing causes. Without this, human factors and management system failures might be missed, resulting in limited benefits. Resource limitations may be important if too many incidents are selected for detailed review and sharing. Legal concerns may also hamper the reporting and sharing of important human factors issues.

The following case history illustrates the consequences of inadequate internal sharing:

During a routine start-up following a maintenance turnaround, the cover of the last reactor in a series abrubtly disengaged as the vessel pressure approached its normal operating pressure of 2,500 psig. The resulting depressurization projected the shell of the reactor downward about 4 feet, into the concrete floor, while launching the reactor cover and internals up and out into open areas. A vapor cloud of hydrogen and ammonia was released, which ignited outside the reactor. Damage to the process unit was extensive and windows were broken in nearby buildings on site. Investigation revealed that the 9-inch threaded retainer, which should hold the reactor head in place, was only engaged to ½-inch at the time of release. Earlier, the start-up process had been interrupted when leaks were found in the adjacent reactor. It had been necessary to de-pressurize and lock-out the reactors for repairs to the head of the leaking reactor. Apparently, the system actuating the retainer on the last reactor was inadvertently and unknowingly activated during maintenance work on the adjacent reactor. The valves activating the hydraulics for opening and closing the head retainers were not close to, nor in line of sight from the other reactors. There were no serious injuries but the cost of the incident, not including business interruption, was in the range of $250,000 to $1,000,000.

One of the key lessons learned was that the actions of the control system on the equipment must be directly observable or positively indicated at the time of operation to ensure that the proper hardware is being activated. Unknown to personnel at the site, a separate locking device, which would have prevented the incident, was in use at a sister plant. The sister plant had developed this device after experiencing a near miss that could have resulted in a similar accident. This event and the corrective action was not communicated throughout the company.

Sharing internally is obviously very important but sharing externally is also important. There are a number of publications and symposia that encourage the reporting of incidents and sharing of results. The U.S. Chemical Safety Board (CSB) issues public reports for every incident it investigates—many management system and human factors issues are presented and addressed in these reports. Similarly, the UK HSE investigates serious incidents and openly publishes an extensive analysis of the results. Obviously, both the U.S. and U.K. believe that the open sharing of incident information is very important.

There are many publications that include incident information. These include:

Process Safety Progress, published by AICHE.

Journal of Loss Prevention in the Process Industries, published by Elsevier.

Annual symposia proceedings of

AICHE Loss Prevention Symposium

AICHE Ammonia Plant Symposium

AICHE CCPS Near Miss Incident Database (for CCPS members only)

AICHE CCPS Process Safety Beacon

U.S. Chemical Safety Board (CSB) Database

Examples of these publications are provided by CCPS (2000), CSB (2004), and CCPS (2005).

Most of these sources rely on reports and/or papers provided by companies on their own incidents. This normally requires a significant effort and much legal scrutiny, but many companies see this as a means to help the process industry avoid future accidents. Many companies are reluctant to provide details for these types of deficiencies unless the incident was only a near miss.

To accomplish the effective sharing of results, management systems should be created to review lessons learned (both successes and failures), to develop recommendations on improvements, and to provide a mechanism to ensure they are implemented properly. This is the "Plan, Do, Check, Act" model of Figure 3-1. Anytime a new program is implemented, it is critical that its effectiveness is measured and changes or modifications made to make it even better.

The objective is to thoroughly entrench human factors programs and make them an important part of the culture of the company. All programs will require minor adjustment from time to time and occasionally a major update will be necessary to reinforce and reinvigorate the principles and requirements.

Useful information on human factors can come from a wide variety of sources— near miss incidents, job safety analyses, task safety observations, and others. The question is how to obtain this information and to use it to improve performance.

In the case of an existing incident investigation program, measurements can be made of the human factors contributors. The results can then be analyzed to determine where weaknesses exist and improvements made.

This program will result in the creation of new procedures and standards—these procedures and standards then protect against similar future incidents. This is usual-

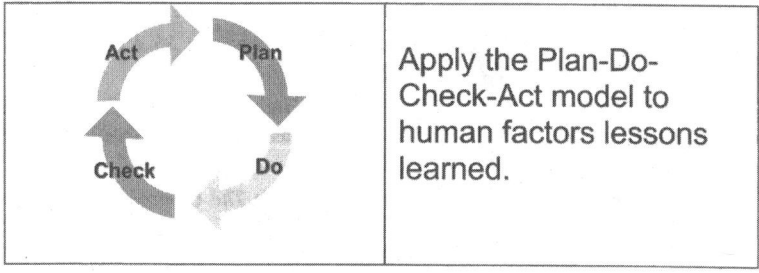

Apply the Plan-Do-Check-Act model to human factors lessons learned.

Figure 3-1: The "Plan-Do-Check-Act" model.

ly effective for a time period since the basis for the procedure or standard is obvious to everyone. However, the challenge to the company is to ensure that the procedure or standard is not modified or relaxed with time because the basis for the original procedure or standard is usually forgotten. Otherwise, the incident(s) may occur again. Part of the continuous improvement process is to ensure that a company does not lose the lessons of the past. Retention of corporate memory is a problem and this subject represents a number of serious human factors issues that become more and more critical as organizations downsize.

So what is required to make the best use of lessons learned with respect to human factors in your own organization? The following list details some of these requirements:

1. Ensure human factors are an important part of lessons learned, including final recommendations and corrective actions.
2. Share human factors lessons learned within the process, facility, organization, and industrial community.
3. Measure and confirm that the application of lessons learned is effective.
4. Build on successes and integrate human factors lessons learned best practices throughout work processes and procedures, company standards and the culture of the organization.
5. Provide feedback to workers.
6. Share human factors information throughout your organization and industry to support continuous improvement.

3.5 REFERENCES

CCPS (2000), *International Conference and Workshop on Process Industry Incidents* (New York: Center for Chemical Process Safety).

CSB (2004) Testimony of Carolyn Merritt, Chairman of CSB, on hearing on "Hazard Communication in the Workplace," March 25, 2004, (Washington DC: U.S. Chemical Safety Board).

CCPS (2005), "Process Safety Incident Database" (New York: Center for Chemical Process Safety).

ExxonMobil (2000) "Human Factors: A Guide to Action," (Irving, TX: Exxon-Mobil).

HSE (2004), *Research Report 194, Human Factors Assessment Model Validation Study* (Glasgow: UK Health and Safety Executive).

OSHA PSM (1992), "Process Safety Management of Highly Hazardous Chemicals; Explosives and Blasting Agents," 29 CFR Part 1910 (Washington, DC: Occupational Safety and Health Administration).

UK COMAH (1999), Statutory Instrument 1999 No. 743: *The Control of Major Accident Hazards Regulations* (The Stationary Office, Norwich, UK).

HUMAN FACTORS TOOL KIT
Facilities and Equipment

This part of the book introduces key topics on human factors identified by the CCPS Human Factors subcommittee. A broad range of human factors topics are covered in a concise and systematic format. Each of the topics is covered by a brief introduction with a focus on relevant human factors principles. This is followed by available tools and industry applications. Where appropriate, graphics are used to reinforce concepts and references are provided for further information. Topics with potential regulatory implications (e.g., hazard analysis and OSHA PSM) are noted as such. Readers may go directly to their topics of interest, as each major chapter in this part is written to serve as a stand-alone description of the topic. The practical approaches to human factors presented in this part will allow plant personnel to directly apply human factors topics of interest to improve process safety performance.

CHAPTER 4

Process Equipment Design

4.1 INTRODUCTION

People operate and maintain processes by interacting with process equipment. Process equipment includes displays, alarms, controls, computers, manual equipment, and personal protective equipment. Human factors issues for process equipment relate to how people interact with and use equipment. The characteristics of the equipment may increase the likelihood of human failures when it is used by people. These human factors issues are addressed by studying the match between the attributes of people and those of the equipment. Optimizing this match helps improve process safety.

Key attributes of people include:

- Physical capabilities such as strength, vision, hearing, and size.
- Intellectual capabilities such as understanding of the process, diagnosing problems, attention span, and speed of thought.
- Skills such as use of tools and reading level; knowledge such as that of process hazards and abnormal operations.
- Behavior such as responsibility and motivation.
- Education and training.
- Communication skills.

Key attributes of equipment include location, accessibility, size, shape, color, orientation, and so on. Process equipment designs should consider capabilities of people such as physical (e.g. strength), psychological (e.g. speed of processing information) and social (e.g. communication abilities).

The diversity of equipment used in processes, their characteristics, and those of people, result in a wide variety of human factors design issues. These are generally well-known. The challenge is to apply the existing knowledge and account for synergistic effects resulting from people and equipment functioning as parts of complex systems.

Fundamental human factors design principles can be defined for equipment and include:

- *Suitability*—meets requirements for the task.
- *Simplicity*—avoids unnecessary complexity.

- *Identifiability*—easily found and recognized by personnel.
- *Accessibility*—easily reached.
- *Detectibility*—easily seen or heard.
- *Availability*—there when needed.
- *Logic and consistency*—organization, arrangement, operation, etc. makes sense.
- *Flexibility*—adjustable to the user.
- *Conformity with user expectations*—operation matches user training and previous experience.

Adherence to these principles should be balanced and tailored to specific situations.

Human interaction with process equipment is affected by the physical, social, and organizational environments in which it occurs. Consequently, issues identified in other sections of this book should also be addressed when considering equipment design. Furthermore, human-equipment interaction should be addressed from an overall systems perspective that considers the entirety of the process and all the people who are part of it. Human factors for alarms and visual display units are covered in detail in Chapter 5 on Process Control Systems and Chapter 9 on the Human/Computer Interface.

Figure 4-1 shows a good application of colors to improve the readability of a pressure gauge.

Figure 4-1: The application of color to improve the readability of a pressure gauge.

An operator used a piece of pipe as a "helper" or "persuader" to open a stiff drain valve on a tank. The valve was infrequently used and not the correct type for this service. He applied too much pressure and broke off the valve handle with the valve in the open position. The entire contents of the tank drained into the dike surrounding the tank.

4.2 TOOLS

Human factors design principles for equipment are well established. There are established texts on the subject (Sanders and McCormick, 1993; Anon., 1986). Various standards also exist. The International Organization for Standardization (ISO) has published a standard on ergonomic requirements for the design of displays and control actuators (ISO 9355-1, 1999). Part 1, Human Interactions with Displays and Control Actuators specifies general principles. Control actuators are defined as the part of the control system that is actuated by the operator. Some guidance is provided on methods that can be used to achieve the principles described. Part 2, Displays, provides guidance on the selection, design and location of displays to avoid potential ergonomic hazards associated with their use (ISO 9355-2, 1999). It specifies ergonomics requirements and covers visual, audible and tactile displays. Part 3, Control Actuators and Part 4, Location and Arrangement of Displays and Control Actuators are under development.

The UK Health and Safety Executive has provided guidelines for data presentation in displays and the selection of controls (HSE, 2004).

The UK Ministry of Defence has an extensive set of standards for the human factors design of equipment (Def Stan 00-25, 1987; 1997a; 1997b; 1992a; 1992b; 1997c; 1996a; 1989a; 1991; 1992c; 1988; 1989b, 1996b) and there are also US military standards relating to human factors design (MIL-STD-1472D, 1989; MIL-HDBK-759A, 1981). Detailed guidance is also available from the nuclear industry (NUREG-0700, 2002; ANSI/IEEE 845, 1988; ANSI/IEEE 1023, 1988; NUREG CR-4227, 1985). Information in these standards can be adapted to the process industries. However, the guidance is highly detailed in many cases and may act as an obstacle to process industry companies beginning to adopt human factors design principles and guidelines for equipment. A simple checklist has been suggested for equipment design in the process industries (Attwood, et al., 2004). The human factors design principles listed in section 20.1 provides a simple guide when designing or modifying equipment.

4.3 REFERENCES

Anon. (1983; 1986), "Ergonomic Design for People at Work," Volumes 1 and 2, Eastman Kodak Company (New York: Van Nostrand Reinhold).

ANSI/IEEE 845 (1988), "Guide to Evaluation of Man—Machine Performance in Nuclear Power Generating Stations, Control Rooms, and Other Peripheries" (Washington, DC: American National Standards Institute).

ANSI/IEEE 1023 (1988), Guide for the Application of Human Factors Engineering to Systems, Equipment and Facilities of Nuclear Power Generating Stations (Washington, DC: American National Standards Institute).

Attwood, D. A., Deeb, J. M. and Danz-Reece, M. E. (2004), "Ergonomic Solutions for the Process Industries" (Amsterdam: Elsevier).

Def Stan 00-25, "Human Factors for Designers of Equipment" (Glasgow: UK Ministry of Defence).

1987:   Part 1: Introduction

1988:   Part 11: Design for Maintainability

1989a: Part 8: Auditory Information

1989b: Part 12: Systems

1991:   Part 9: Voice Communication

1992a: Part 4: Workplace Design

1992b: Part 5: Stresses and Hazards

1992c: Part 10: Controls

1996a: Part 7: Visual Displays

1996b: Part 13: Human Computer Interaction

1997a: Part 2: Body Size

1997b: Part 3: Body Strength and Stamina

1997c: Part 6: Vision and Lighting

HSE (2004) http://www.hse.gov.uk/comah/sragtech/techmeascontrol.htm (London: UK Health and Safety Executive).

ISO 9355-1 (1999a), "Ergonomic requirements for the design of displays and control actuators—Part 1: Human interactions with displays and control actuators" (Geneva, Switzerland: International Organization for Standardization).

ISO 9355-2 (1999b), "Ergonomic requirements for the design of displays and control actuators—Part 2: Displays" (Geneva, Switzerland: International Organization for Standardization).

MIL-HDBK-759A (1981), "Military Handbook, Human Factors, Engineering Designs for Army Material" (Washington, DC: U.S. Department of Defense).

MIL-STD-1472D (1989), "Human Engineering Design Criteria for Military Systems, Equipment and Facilities" (Washington, DC: U.S. Department of Defense).

NUREG/CR-4227 (1985), "Human Engineering Guidelines for the Evaluation and Assessment of Visual Display Units" (Washington, DC: U.S. Nuclear Regulatory Commission).

NUREG-0700 (2002), "Human-System Interface Design Review Guidelines," Rev. 2 (Washington, DC: U.S. Nuclear Regulatory Commission).

Sanders, M. S. and McCormick, E. J. (1993), "Human Factors in Engineering and Design," 7th ed. (New York: McGraw-Hill).

4.4   ADDITIONAL REFERENCES

BS EN 894-1 (1997), "Safety of machinery. Ergonomics requirements for the design of displays and control actuators. Part 1. General principles for human interactions with displays and control actuators" (London: British Standards Institution).

BS EN 894-2 (1997), "Safety of machinery. Ergonomics requirements for the design of displays and control actuators. Part 2. Displays" (London: British Standards Institution).

BS EN 60073 (1997), "Basic and safety principles for man-machine interface, marking and identification. Coding principles for indication devices and actuators" (London: British Standards Institution).

BS 3693(1992), "Recommendations for the design of scales and indexes on analogue indicating instruments" (London: British Standards Institution).

BSR/HFES 100 (2002), "Human Factors Engineering of Computer Workstations, Draft Standard for Trial Use" (Santa Monica, CA: Human Factors and Ergonomics Society).

HSE (1992) "L26, Display Screen Equipment Work—Guidance on Regulations" (London: UK Health and Safety Executive).

HSE (1998), "The Management of Alarm Systems," CRR 166/1998 (London, UK Health and Safety Executive).

IEC 61310-1 Ed. 1.0 b (1995), "Safety of machinery—Indication, marking and actuation—Part 1: Requirements for visual, auditory and tactile signals" (Geneva: International Electrotechnical Commission).

IEC 61310-2 Ed. 1.0 b (1995), "Safety of machinery—Indication, marking and actuation—Part 2: Requirements for marking" (Geneva: International Electrotechnical Commission).

IEC 61310-3 Ed. 1.0 b (1999), "Safety of machinery—Indication, marking and actuation—Part 3: Requirements for the location and operation of actuators" (Geneva: International Electrotechnical Commission).

ISO 13849-1 (1999), "Safety of machinery—Safety-related parts of control systems—Part 1: General principles for design" (Geneva, Switzerland: International Organization for Standardization).

ISO 13849-2 (2003), "Safety of machinery—Safety-related parts of control systems—Part 2: Validation" (Geneva, Switzerland: International Organization for Standardization).

ISO/TR 13849-100 (2000), "Safety of machinery—Safety-related parts of control systems—Part 100: Guidelines for the use and application of ISO 13849-1" (Geneva, Switzerland: International Organization for Standardization).

ISO 12100-1 (2003) "Safety of machinery—Basic concepts, general principles for design—Part 1: Basic terminology, methodology" (Geneva, Switzerland: International Organization for Standardization).

ISO 12100-2 (2003), "Safety of machinery—Basic concepts, general principles for design—Part 2: Technical principles" (Geneva, Switzerland: International Organization for Standardization).

ISO 13850 (1996), "Safety of machinery—Emergency stop—Principles for design" (Geneva, Switzerland: International Organization for Standardization).

ISO 7731 (2003), "Ergonomics—Danger signals for public and work areas—Auditory danger signals" (Geneva: International Organization for Standardization).

CHAPTER 5

Process Control Systems

5.1 INTRODUCTION

The process control system links the human operator to the process equipment. The designer must decide which functions to allocate to the process control system and which to allocate to the operator. The current trend is to automate the control functions, and that trend has accelerated as computers, sensors, and data networks have become more capable and reliable. This has reduced or eliminated much repetitive manual work and facilitated both productivity and quality gains as fewer people were required to operate more/larger processes.

In normal operation, the dominant issue for process control is designing the system to minimize the potential for human error. A well-designed process control system can check the validity of operator inputs based on the state of the system. For example, if 3,000 liters have already been added to a 4,000-liter reactor, the control system can reject an operator input calling for the addition of 2,000 liters of solvent. The control system can also dampen extreme control moves so a valve does not slam closed instantly when the operator changes the setpoint to 0. Human factors considerations in process control will help operators avoid upsets and help operators diagnose and respond to those upsets that do occur.

However, to provide these benefits, the control system itself has become increasingly complex. Redundancy of the physical elements (sensors, logic solvers, displays, etc.) is necessary to achieve the required reliability, but the theoretical benefit of such redundancy is severely limited by its ultimate reliance on—humans! Humans can introduce common cause failures by miscalibrating multiple similar instruments, misprogramming the software, leaving the system in test mode, and so on. The more complex the control system, the more difficult it is for workers to maintain the system and to diagnose operational upsets. In highly automated systems, the operator must be an expert troubleshooter, ready to step in and take control in unusual situations. This challenges the process designer to provide the human with the necessary information to recognize, diagnose, and effectively respond to upset conditions.

5.2 ISSUES/EXAMPLES

The single biggest issue in process control is abnormal situation management. The operator in a control room is deprived of much of the feedback that immerses a hands-on operator (sounds, smells, vibrations, etc.). Thus, process designers try to compensate by providing alarms that call the operator's attention to unusual plant conditions. This approach is fine as long as there are only a few intermittent alarms. However, if there is a major upset, dozens (sometimes hundreds) of alarms "flood" the operator with information that is impossible to comprehend. If the operator is expected to diagnose the situation and prevent a unit shutdown, then an effective system of alarm prioritization is required. Low priority alarms must be suppressed so the operator can focus attention on the status of critical equipment.

In one facility, a large process upset overloaded the computer, resulting in the computer not updating the displays. The operator was unable to determine the state of the process from the computer displays.

A closely related issue is the management of nuisance and perpetual alarms. These types of alarms typically result when alarm setpoints are poorly chosen, when the equipment and/or control system is poorly maintained, or when the alarm logic takes no account of the equipment status (e.g., active alarms on idle equipment). Regardless, such alarms train the operators to ignore them or to aggressively disable them (with rags stuffed in speakers, toothpicks wedged in acknowledge buttons, wires disconnected, etc.), which inevitably leads to human errors.

In one facility, operators ignored what they believed to be nuisance alarms warning of a runaway reaction. The ensuing explosion and fire killed four workers and devastated the facility.

The process control system must be designed to provide enough information for the operator to quickly diagnose the cause of the upset and respond to it. Simply calling attention to the situation with an alarm is not enough. This requires that key operating parameters be displayed in overview fashion (analog and trend displays are usually preferred) and that operators be able to quickly navigate to detailed displays as needed. The displays must be updated quickly to show the operators immediately that their control moves are having an effect; otherwise operators tend to overshoot the desired setpoint or toggle digital inputs repeatedly into an unknown state.

A management issue in process control is whether to allow operators to "force" an input into a desired state. The capability for an operator to override the automated system is often desirable when the system is upset and the operator can confirm that the problem is, for example, simply a bad sensor. Another situation occurs when the operator may need to respond to a situation that was not anticipated by the designers. However, if operators use the overrides repeatedly because equipment failures remain uncorrected, disaster is sure to result. Overrides also allow operators to bypass carefully planned system responses with spur-of-the-moment actions. So, if overrides are allowed, there must be management controls in place to ensure that overrides are not abused.

5.3 TOOLS

The tools used to gauge operator workload, as discussed in Chapter 16, may be used to evaluate the process control system's compatibility with operator needs.

5.4 REFERENCES

Bransby, M.L. and Jenkinson, J. (1998), "CRR166: The Management of Alarm Systems CRR 166" (Sudbury, UK: Health and Safety Executive).

EEMUA (1999), "Alarm Systems, a Guide to Design, Management, and Procurement No. 191" (London: Engineering Equipment and Material Users Association).

HSE (2000), "Information Sheet: Better Alarm Handling" (Sudbury, UK: Health and Safety Executive).

HSE (1999), "HSG48: Reducing Error and Influencing Behaviour" (Sudbury, UK: Health and Safety Executive).

ISA (1985), "Human Engineering for Control Centers: ISA Standard RP60.3" (Research Triangle Park, NC: The Instrumentation, Systems, and Automation Society).

Kletz, T., Chung, P., Broomfield, E., and Shen-Orr, C. (1995). *Computer Control and Human Error* (Houston, TX: Gulf Professional Publishing).

CHAPTER 6

Control Center Design

6.1 INTRODUCTION

Control centers should be organized to provide personnel with a working environment that helps ensure safety and operability. Effective human factors control center design can reduce the likelihood of human errors in the control center and the likelihood that control center personnel will be exposed to ergonomic hazards.

Control centers are often part of a complex of functionally-related rooms which house supporting functions for the control center. This complex may include offices, equipment rooms, rest areas, and training rooms, and is sometimes called the *control suite.* Human factors are important for the entire control suite.

Human factors design for control rooms deals with how people interact with the room and its contents. Layout must be addressed. It includes access and egress, lines of sight, impact on communications (both verbal and non-verbal), allocation of responsibility and the requirements for supervision, range of staffing levels, workstation separation, personnel movement around the control room, and so on. Other topics in control room design that should be addressed include

Visual Display Units (VDUs) and panels for displays and controls (see Chapter 9 on the Human/Computer Interface), alarms (see Chapter 5 on Process Control Systems), workstation anthropometry including reach, seating, adjustment, and posture (see Chapter 8 on Facilities and Workstation Design), communications since control room personnel must be able to communicate effectively with each other and other plant personnel (see Chapter 13 on Communications), environmental issues such as temperature, humidity, air quality, lighting, noise, and so on (see Chapter 15 on Environmental Factors) and maintenance for issues such as access, labeling, and so on (see Chapter 23 on Maintenance).

Control room design should address both normal and abnormal operations and 24-hour operation where that is employed. It is also important to design control rooms consistently when there is more than one control room at a plant.

A control room staffed by a single operator was located on the third floor of a building in the middle of the process equipment. While the operator was working on equipment on the second floor of the building a vapor release occurred. The operator was prevented from returning to the control room and a larger release occurred than would have otherwise been the case.

6.2 TOOLS

Ergonomic design of control centers is addressed by ISO 11064 (ISO 11064, 2000a; 2000b; 2002, 2004a; 2004b; 2005; 2006). This standard covers design principles, control room arrangements and layout, workstations, displays, controls, interactions, temperature, lighting, acoustics, ventilation, and evaluation. Designers can follow this standard for new control rooms and for upgrades and modifications to existing ones, especially where there are known problems.

Other guidance is also available (EEMUA, 2002; WIB, 1998a; WIB, 1998b; WIB, 1997; ISA RP60.1, 1990; ISA RP60.2, 1995; ISA RP60.3, 1985; ISA RP60.4, 1990). EEMUA Publication 201 (EEMUA, 2002) is a guide for sites moving to distributed control systems (DCSs) and centralized control rooms. It contains several examples of how companies have approached such projects including graphics design.

The Norwegian Petroleum Directorate (NPD) has developed an audit method for evaluating human factors considerations for control rooms (NPD, 2003). It is intended as a verification and validation tool and not as a guideline for design.

Figure 6-1 shows a good control room layout. The control panels and displays are well-organized and accessible. The lighting and environmental conditions are also good.

Figure 6-1: An example of a control room layout.

6.3 REFERENCES

EEMUA (2002), "Process plant control desks utilizing human-computer interface: A guide to design, operational and human interface issues," Publication 201 (London: Engineering Equipment & Materials Users Association).

ISA-RP60.1(1990), "Control Center Facilities" (Research Triangle Park, NC: The Instrumentation, Systems and Automation Society).

ISA-RP60.2 (1995), "Control Center Design Guide and Terminology" (Research Triangle Park, NC: The Instrumentation, Systems and Automation Society).

ISA-RP60.3 (1985), "Human Engineering for Control Centers" (Research Triangle Park, NC: The Instrumentation, Systems and Automation Society).

ISA-RP60.4 (1990), "Documentation for Control Centers" (Research Triangle Park, NC: The Instrumentation, Systems and Automation Society).

ISO 11064, "Ergonomic design of control centers" (Geneva, Switzerland: International Organization for Standardization).

1999, 2002:	Part 3: Control room layout
2000:	Part 1: Principles for the design of control centers
2000:	Part 2: Principles for the arrangement of control suites
2004:	Part 4: Layout and Dimensions of Workstations
2004:	Part 5: Human-System Interfaces
2005:	Part 6: Environmental Requirements for Control Centers
2006:	Part 7: Principles for the Evaluation of Control Centers

NPD (2003), "HF Assessment Method for Control Rooms" (Stavanger, Norway: Norwegian Petroleum Directorate), www.npd.no

WIB (1997), "Ergonomics in Process Control Rooms, Part 3: The Analyses," Report. No. M 2657 X 98 (The Hague, Netherlands: International Instrument Users' Association)

WIB (1998a), "Ergonomics in Process Control Rooms, Part 1: Engineering Guideline."

ISO 11064-1 (2000), "Ergonomic design of control centers," Report. No. M 2655 X 98 (The Hague, Netherlands: International Instrument Users' Association).

WIB (1998b), "Ergonomics in Process Control Rooms, Part 2: Design Guideline," Report No. M 2656 X 98 (The Hague, Netherlands: International Instrument Users' Association).

6.4 ADDITIONAL REFERENCES

ISO 6385 (2004), "Ergonomic Principles in the Design of Work Systems" (Geneva, Switzerland: International Organization for Standardization).

ISO 9921(2003), "Ergonomics—Assessment of Speech Communication" (Geneva, Switzerland: International Organization for Standardization).

Ivergard, T. (1989), "Handbook of Control Room Design and Ergonomics, (Boca Raton, FL: CRC Press).

Kincade, R. G. and Anderson, J. (1984), EPRI—NP3659: "Human Factors Guide for Nuclear Power Plant Control Room Development" (Palo Alto, CA: Electric Power Research Institute).

Noyes, J. and Bransby, M. (eds.) (2002), "People in Control: Human Factors in Control Room Design" Control Engineering Series, Vol. 60, (London: Institution of Electrical Engineers).

CHAPTER 7

Remote Operations

7.1 INTRODUCTION

Remote operation of process plants is defined as the absence of a direct interface between operators and process equipment, usually accomplished through the use of computers with a control room. Alternatively, remote operation can be defined as the presence of some form of barrier between the process and the operator that requires the indirect transmission of process information (HSE, 2002).

In practice, various types of remote operations may be encountered, including the following and combinations thereof:

- A process is physically separate from the control room, but the operators monitor and control the process on a continuous basis (satellite operation).
- A process operates autonomously, and the automated control system requests attention from operators as needed (unmanned).
- A process is controlled by local operators but supervisory operators may take control, as needed (remote monitoring and/or control).

Remote operations can be located proximately or at considerable distance from the control room. Proximate operations usually involve both field operators (FO's) working out in the process and control room operators (CRO's) working in the control room. Control room operators may also act as field operators in some cases. Distant operations may also involve both types of operators. However, field operators are less likely to be present at distant operations on a full-time basis—the operations are likely to be fully automated and unmanned with attention from operators provided only when needed.

Remote operations raise significant issues related to safety and operability—control room operators may be less likely to observe problems owing to their lack of a direct and continual interface with the process. Key human factors issues for remote operation are:

- *Communications.* There is usually less face-to-face communication between field and control room operators compared to manually operated processes. Radios, phones, public address systems, etc. may be used more and their reliability is an issue. Shift handovers may involve both FO's and CRO's and require more care. Operating multiple process units in a single, centralized control room may increase communication between operators of the different units.

- *Nature of tasks.* Tasks are more likely to involve the FO and CRO working together, with requirements for cooperation and communication.
- *Provision of operating information.* Operators rely on visual display units (VDUs) rather than local analog gauges and display panels. Analog displays show all information at all times whereas VDUs may provide more information but less is visible at any given time. Also, the operator may lose some spatial awareness provided by display panels.
- *Detection of process problems.* There are fewer opportunities for direct perception (sight, sound, smell).
- *Level of distraction.* Remotely located control rooms may have fewer visitors and consequently fewer distractions. Conversely, centralization of the operation of multiple units in a single control room may create a more distracting environment.
- *Travel time.* Depending on the location of the control room, there may be significant travel time for CRO's to reach the field and FO's to reach the control room. Consequently, less time will be spent in the plant or the control room.
- *Response times.* Generally, remote operations will cause response times for detecting and dealing with malfunctioning equipment to increase.
- *Plant excursions by the CROs.* Plant excursions must be planned for remote operations. Operators may not be able to easily and quickly reach the plant to reset trips and start/stop equipment.
- *Control room is a focal point.* The control room of a remotely operated plant becomes the focal point of operations. CROs are less likely to visit the field. Managers, supervisors, and engineers are less likely to spend time in the field where they may observe housekeeping and plant condition issues. Their presence in and around the control room may also irritate or intimidate the operators, and/or make them less likely to act quickly, decisively, and autonomously.
- *Socio-technical issues.* Different classes of operators and inequitable work environments may be created. There is usually less personal interaction between operators. Benefits to operators of remote operation may include more interesting work, improved work environment, higher compensation, and improved stature and career prospects. Costs may include less challenging work, more sedentary work, loss of job security and job satisfaction, fear of change, and the need to learn new ways and adapt to new systems.
- *Lone workers.* Operators may visit remotely operated facilities alone and be at greater risk if they are exposed to hazards since no one else is present to observe their need for assistance and help is further away.

Change to remote operations is often accompanied by organizational changes such as downsizing, centralization and multi-tasking (see Chapter 25 on Management of Change). The synergistic effects of such changes should be considered holistically.

Figure 7-1 shows a control room for monitoring and controlling a number of remote facilities. The computer screens are identical to what an operator would see on a similar screen in the control room at the remote facility. This company also has remote/satellite plants, where there are no operators (normally), but a "manager" that lives and works somewhere else. When these plants have problems or require adjustment, the "manager" dials-in via laptop and makes changes (can be as radical as start-up/shutdown, in rare cases) just as though he/she were in the control room at the plant.

An operator visiting an unmanned plant fell from a ladder. He dropped his radio during the fall and could not retrieve it owing to his injuries. Several hours passed before his absence was noticed.

7.2 TOOLS

There are no tools currently available that deal specifically with human factors for remote operations. A research report discusses some of the issues involved with

Figure 7-1: A control room for operation of a number of remote facilities. Photo courtesy of Air Products and Chemicals.

Table 7-1: Basic human factors guidelines for remote operations

---

- Establish communication protocols (see Chapter 13 on Communications).
- Define tasks and write procedures to delineate the responsibilities of inside and outside operators.
- Provide appropriate training for remote operations.
- Provide appropriate rules and monitoring.
- Qualify operators for remote operations.
- Address emergency operations, including communications.
- Rotate field and control room operators to providing a better understanding of their respective jobs and improved communications.
- Design the control system to insure the displays provide sufficient and appropriate information to allow operators to determine what is happening in the process.
- Install CCTV, as appropriate, to allow visual monitoring of the process.
- Schedule and perform operator walk-throughs, as appropriate. Operators can identify minor leaks and other problems that may be leading indicators of pending upsets or incidents. Require the manual completion of logs to promote the discipline of walking around the plant.
- Design control rooms to minimize distractions.
- Consider operator travel time in locating control rooms.
- If a remote operation is intermittently manned by a single operator, utilize an "operator-down" device to protect the operator.

---

changing to remote from manual operations (HSE, 2002). A number of the issues are addressed by other sections of this book. Table 7-1 provides some basic guidelines.

A control system at a facility stopped functioning on Christmas Eve. The control system program had been corrupted and "self-deleted" itself, causing the plant to trip. Since the program was deleted, they could not restart the plant. The process controls engineer logged in to the corporate computer from home (in another continent) at 4:00 am and reloaded the entire program. The plant was back running and produced product later in the same day.

7.3 REFERENCE

HSE (2002), Human Factors Aspects of Remote Operation in Process Plants, UK Health and Safety Executive, Contract Research Report 432 (London: U.K. Health and Safety Executive).

CHAPTER 8

Facilities and Workstation Design

8.1 INTRODUCTION

In this chapter, the workplace is defined as "*the physical area where a person performs tasks.*" The workplace may include physical fixtures such as furniture, equipment, hallways, stairs, vehicles, and displays and will be affected by environmental variables such as lighting, temperature, and noise. A workstation is defined as "*a location where the operator may spend only a portion of the working shift.*" Clearly, workstations are a subset of the workplace. An operator may travel between and work at several different workstations in the workplace.

A good workplace and workstation design is based on biomechanical, physiological, and psychological requirements of the user. Body stress can result from manual material handling, constrained posture, frequent movement between tasks, and frequent repetitions. Poor work flow and layout can cause unnecessary fatigue even as production processes become more automated. For example, a person unloading bales from an automated conveyor system may not be able to keep up with the system. The operator should be able to control and adjust the speed.

The following are some general principles in work place and workstation design that were taken from Attwood et al. (2003). The reader should note that specifications exist for most of the general statements. However, specific dimensions and forces are beyond the scope of this book. The interested reader should read the many sources available if more detail is required.

- Design workplaces to accommodate the extremes of the user population.
- Design workplaces to adjust to the characteristics of the user population (Figure 8-1).
- Design equipment to be physically accessible (Figure 8-2).
- Avoid forcing a joint beyond its natural range of movement, especially while applying force, or holding posture (Figure 8-3). The solution to the valve posture issue shown in Figure 8-3, is to specify the preferred location and orientation of valves as demonstrated in Figure 8-4.
- Avoid holding tensed muscles in fixed positions for long periods.
- Locate frequently accessed items within easy reach from the working position.
- Locate work with hands at approximately elbow height depending on the task.

Figure 8-1: Adjustable height workstation (Photograph courtesy Evans Consoles Inc. Calgary AB, Canada).

- Minimize highly repetitive tasks.
- Avoid frequent and repetitive high contact force.
- Provide specialized tools to reduce body stress (Figure 8-5).
- Arrange computer-based workstations and seating according to accepted human factors standards (BSR/HFES, 2002).
- Ensure that the working environment, especially temperature and lighting, is properly designed
- Schedule proper rest breaks depending on the work effort required by the task.

For sitting tasks all items should be within reach, the hands should work no more than 6 inches above the work surface, and the weight of objects handled should not exceed 10 pounds (Attwood, 1996). Special adjustments may be required for fine assembly, writing, or precision work.

For standing tasks avoid:

- Frequent high and/or low extended reaches.
- Frequent movement between work stations (several times per minute).
- Work stations with no knee clearance.
- Exerting downward forces.

For combinations of sitting and standing tasks, design elevated workstations and seating that allow workers to move easily between seated and standing positions.

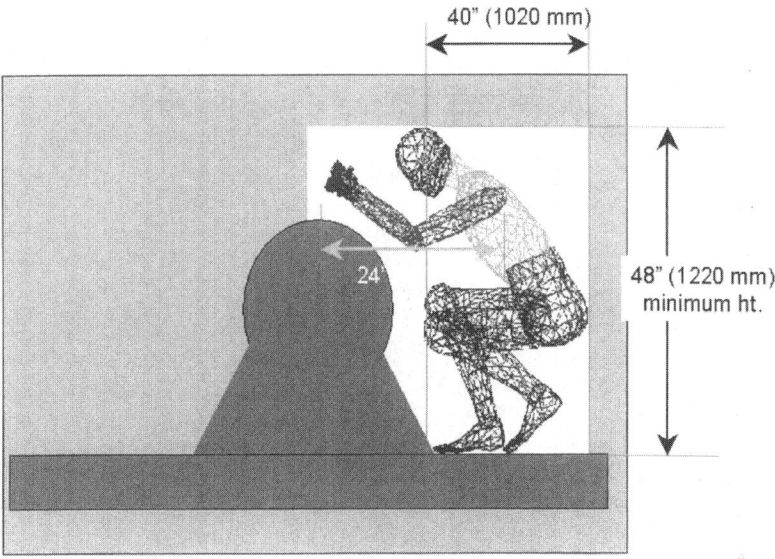

Figure 8-2: Minimum clearance required for a technician to work on a pump from a squatting position. [Reproduced from Attwood et al. (2003).]

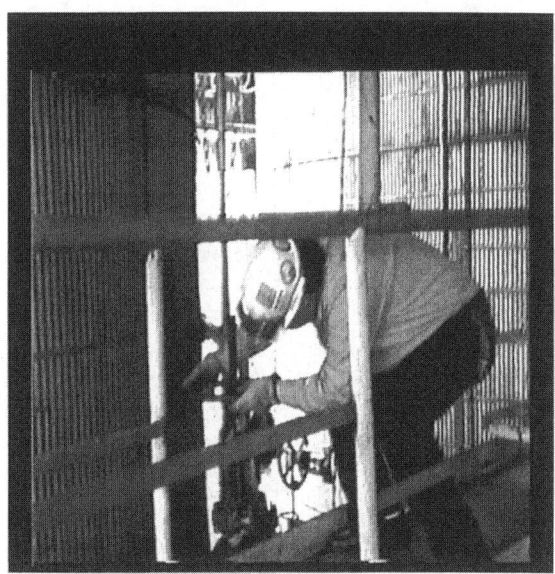

Figure 8-3: Technician in an awkward posture while applying force on a valve wheel.

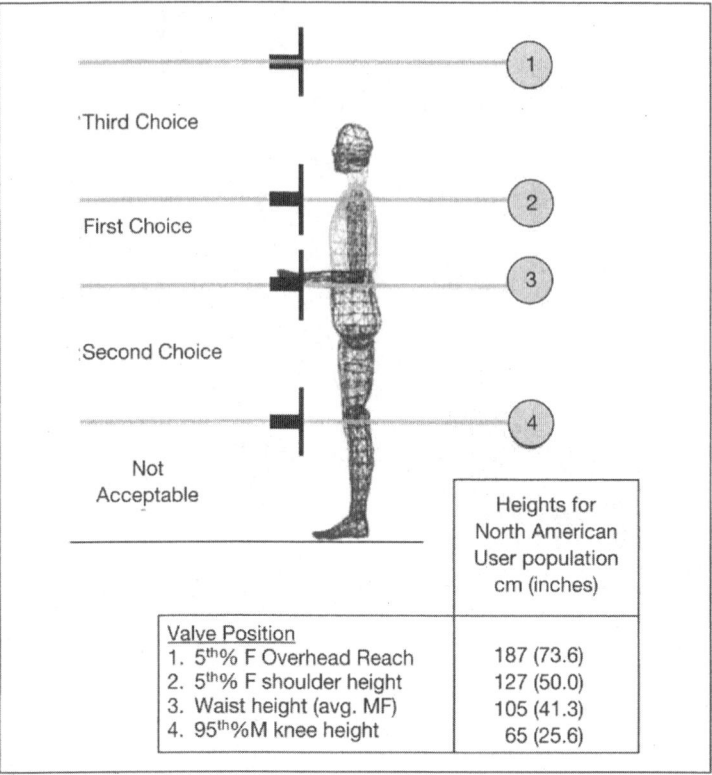

Figure 8-4: Preferred mounting heights of horizontal stem valves.

This category of workstation can offer users the best of both worlds, since they do not suffer the fatigue experienced from standing, yet have more range of motion than is offered by seated workstations alone.

Equipment breakdowns can affect productivity and increase the costs of equipment replacement and repair. Two principles should be followed in this area. First, equipment must be designed for ease of maintenance. It should be provided with adequate access panels, doors, shelves, and drawers so that access is assured to components that frequently wear out and must be replaced. Second, equipment must be installed so it can be maintained properly. Operators must be able to access equipment safely and easily to prevent injury and enhance efficiency.

A pressure gauge was installed in a location that was not easily visible to the operator. A mirror was installed to solve the problem. This resulted in the operator reading the pressure gauge incorrectly since the gauge direction was now reversed!

Figure 8-5: Hydraulic wrench designed to reduce the requirement for operator force at a well head

8.2   TOOLS

The proper design of workplaces can be assessed using several tools that are commonly used by human factors specialists.

Activity Analyses

Activity analyses are used to determine the time spent on each work place task and when each task is performed. The separate activities that make up the task are identified prior to data collection. Then, operator behavior is observed at standard intervals during the task and the activity performed is noted. At the end of the task, the number of times each activity is performed is totaled, then multiplied by the sampling interval to obtain the total time per activity (Eastman Kodak, 1983)

Task Analysis

In its simplest form, a task analysis is a systematic process that is used to:

- produce an ordered list of activities that people perform in a task, and
- identify the human factors issues associated with each activity.

The technique presented in Attwood et al. (2003) requires the analyst to develop a functional description of the task, then develop a list of sequential activities and analyze them. While there are many variations on the analysis, the process typically involves identifying the error potential, the human factors issues and the potential intervention for each activity (Woodson, 1981).

Link Analysis

Link analysis optimizes the location of people and equipment in a workplace. It considers the following principles:

- Services needed by several people should be located in a central location.
- People and equipment should be located to maximize communication between them.

Table 8.1: Checklist for Human Factors Task Assessment. It can be difficult to know when an in-depth Human Factors assessment is required. The following checklist may be of use to the site HF team to help guide their thought process. As with all checklists, this one is not meant to capture all possible HF issues. Its purpose is to identify situations where additional evaluation is required. If any of the questions are answered with "yes," a more in-depth job evaluation should be considered.

Questions	No	Yes	Comments
1. Do workers complain frequently about this or similar tasks/work spaces?			
2. Is turnover on this task excessive?			
3. Is product quality frequently too low on this or similar tasks/work spaces?			
4. Is productivity too low on this task?			
5. Does this task/work space involve the use of new technology, work methods, tools, or work procedures?			
6. Is there a history of possibility of personnel assigned to this task being restricted due to body size, strength, age, gender?			
7. Is the training time unusually long for this or similar tasks?			
8. Do workers frequently try to change the work space/methods for "comfort"?			
9. Are workers frequently away from the work space performing unnecessary activities?			
10. Is the work space used on more than one shift per day?			
11. Is the work space used by more than one individual per day?			
12. Is there a record of cumulative trauma disorders associated with This task/work space or with a similar one?			
13. Is there a record of overexertion associated with this or a similar task?			
14. Is a record of "human error" accident events associated with this or a similar task/work space?			
15. Is there a record of medical cases or numerous visits to medical facilities associated with this or a similar task/work space?			

- Distances between components that are important to each other should be minimized.
- The probability of unnecessary physical interaction between people or equipment should be minimized.

The technique involves identifying the site areas and equipment in the workplace and measuring the distance between them. The analyst then relocates areas and equipment according to the principles above, then recalculates the distances to ensure they have decreased (Attwood et al., 2003).

Checklists

Checklists are most often used to assess the need for a more thorough, in-depth analysis. The checklist provided in Table 8-1 is intended to identify situations in the work process where conditions, including workstation design, are not optimal and where additional attention and study are required.

8.3 REFERENCES

BSR/HFES 100 "Human Factors Engineering of Computer Workstations." Human Factors and Ergonomics Society (HFES), P.O. Box 1369, Santa Monica, CA: 2002.

Attwood, D. A., Deeb, J. M, and Danz-Reece, M. E. (2003) "Ergonomics solutions for the Process Industries" (Burlington, MA: Gulf Professional Publishing).

Attwood, D. (1996) "The Office Relocation Sourcebook" (New York: John Wiley & Sons, Inc.).

Eastman Kodak (1983) "Ergonomic Design for People at Work," vol. 2 (New York: Van Nostrand Reinhold).

Woodson, W. E. (1981). "Human Factors Design Handbook" (New York: McGraw-Hill.

8.4 ADDITIONAL REFERENCES

Sanders, M. S. and McCormick, E. J. (1993), "Human Factors in Engineering and Design" (New York: McGraw-Hill).

Grandjean, E. and Oldroyd, H. (translator) (1988), "Fitting the Task to the Man: A Textbook of Occupational Ergonomics" (New York: Taylor and Francis).

CHAPTER 9

Human/Computer Interface

9.1 INTRODUCTION

The human/computer interface (HCI) deals with how people interact with computer systems with the objective of ensuring that computer system designs are functional, easily operable, efficient, and safe. The HCI has been studied in some detail for business computer systems (ISO 9241-1, 1997; BS 7179, 1990). The key aspect of the HCI for process plants is human interaction with the control system. Most plants today utilize computer systems such as DCSs, SCADA systems and PLCs to control plant processes. Human failures in interacting with control systems may result in loss of control and serious incidents. Therefore, human factors for the HCI are particularly important for process safety.

Human interactions with computers may be physical, perceptual or sensing (visual, and to a lesser extent auditory and tactile), cognitive (using a mental thought process) or diagnostic (problem solving), and require issuing orders or instructions. Human factors aspects involved are:

- Individual human characteristics (skills, knowledge, anthropometrics—human body measurements, etc.).
- Ergonomics of human interactions with computer equipment.
- Human interaction with control software.

The components of computer control systems that may influence the HCI are:

- Human operator(s)
- Visual display devices
- Data entry devices (keyboards, mice, trackballs, touch screens, light pens, etc.)
- Other computer peripherals (printers, data storage devices, etc.)
- Workstations (chairs, desks, tables, etc.)
- Software
- User documentation
- Alarms
- Environment

This section focuses on the interaction of people with the control system software and the screen displays on visual display units (VDUs)—this is a unique HCI issue for process control systems.

Human interaction with control system software involves several types of tasks: monitoring, detection, diagnosis, data entry, and control. All these tasks are subject to human failures including:

- Not reading information correctly.
- Not responding to system prompts or alarms, or responding incorrectly.
- Incorrectly diagnosing situations.
- Incorrectly entering data or not entering data when required.
- Incorrectly issuing commands or instructions, or not issuing them when required.

Human factors influence the likelihood of these failures.

9.2   HUMAN INTERACTIONS WITH CONTROL SYSTEM SOFTWARE

Operators interact with control system software through various types of visual display devices such as consoles, workstations, human-machine interfaces (HMIs), and handheld programming devices. Computer control systems provide information to operators visually and through audible alarms. Visual alarms may appear on computer screens or as separate alarm lights on a panel. Graphical user interfaces are emphasized in modern control systems and sounds are not typically used, other than for alarms. Audio can contribute positively to the HCI and its use may become more common in the future.

Operators respond to information presented to them by display devices. They exercise control over the process through instructions issued to the software running on the control computers. Note that operators may also use closed-circuit video displays to observe the process. These are not addressed in this section.

Display design is especially important since these displays are the principal means by which operators interact with and control the process. This is an issue that should be addressed during the design of new processes but can also be considered in retrofits for existing processes.

Note that there is a distinction between the ergonomic design of the physical display device and that of the software display. The former deals with the suitability of the display device regardless of the software using it. This section addresses human factors issues relating to how the operator interacts with the process control software displays shown on the display device.

Physical interactions of operators with process control software tend to be fairly simple and limited in number. They generally involve the selection of screen objects, entry or ramping of numerical values, and issuing of simple instructions or commands (e.g. open valve, stop pump). Keys issues include:

- Ease of data entry and of issuing instructions.
- Speed with which data can be entered and instructions issued.
- Input error checking and feedback on incorrect entries.
- Feedback on control actions taken (pending, status, complete).
- System response time.
- Clarity of the control scheme.

Other human factors issues for process control systems such as allocation of function, redundancy, feedback, learning time, and performance-shaping factors are addressed in Chapter 5 on Process Control Systems.

> The start buttons for pumps in a process were organized in a block on a touch screen with minimal separation. A new operator with large hands inadvertently activated the wrong pump causing a spill.

9.3 TOOLS

Most existing VDU standards and guidelines have limited applicability to process control. Many have been developed for office applications and focus on the protection of users from musculo-skeletal disorders and other hazards of using visual display terminals ((VDTs) (ISO 9241-1, 1997; BS 7179, 1990). Some of this work can be adapted to control systems and some control-system specific guidance is available (Gilmore, et al., 1989; Invergard, 1989; HSE 60, 1993). Detailed specifications have been developed for some aspects of the HCI, such as the ergonomics of interactions with display screens, but most guidance is performance-based and not prescriptive.

Def Stan 00-25 (1996) describes the application of task and risk analysis methods to human factors issues involved in the analysis, design and development of HCIs for military systems. Human factors design principles and guidelines are presented. MIL-STD-1472D (1989) also provides human factors guidelines for military systems. NUREG-0700(2002) provides detailed human factors design review guidelines for nuclear power plants. These documents contain material that can be adapted to control systems in the process industries.

There is no single comprehensive standard or guideline that applies to VDUs for process control. However, ISO does plan to issue ISO 11064, Part 4, "Visual Displays for Process Control" in their series "Ergonomic Design of Control Centers." There are existing standards on graphical symbols intended for use with process displays (ISA-5.3, 1983; ISA-5.5, 1985). ISA has announced the development of a HMI standard and has also formed the ISA-SP101 committee for this purpose.

Most of the available tools focus on HCI at the design stage. The focus is on un-

Table 9-1: Checklist of key issues for sensing, understanding and using control software displays.

Individual screen design
- Suitable basis for screens—use of familiar metaphors (process structure, process functions, tasks, etc.).
- Legibility of text (font size and type).
- Suitable representations for the presentation of information (numeric value, mimics, tabular, trends, bargraphs, etc.)
- Symbols, text and numerical information presented appropriately.
- Screen layout (appropriate use of white space, emphasis, and fonts)
- Adequate spacing between active areas of touch screens.
- Screens and information readily identifiable (use of titles, labels, icons).
- Display information in a format adapted to the operator.
- Appropriate and relevant information presented.
- Level of detail and amount of information presented (apply parsimony principle—provide only information the operator needs).
- Appropriate and logical location and arrangement of information.
- Frequency of use and importance of information and controls governs their location.
- Screen schematics correlate with actual plant configuration.
- Iconic symbols use text labels.
- Appropriate coding used (shape, color, alpha-numerics).
- Color used appropriately (meet cultural expectations; coding used consistently; recognize and design for color vision deficiency; appropriate choice, use and number of colors; appropriate color combinations).
- Contrast is used to identify and distinguish important items.
- Avoid display clutter.
- Appropriate use of pan and zoom displays.

Page design
- Assignment of information to screens.
- Sequence of presentation of information.
- Appropriate subdivision of the process across screens.
- Rapid, easy and direct navigation/linkages between pages.
- Number of pages available/to be monitored is appropriate to the circumstances.
- Appropriate balance of summary versus detailed displays.

Consistency
- Screen design and layout
- Terminology
- Meet user expectations
    prior experience
    cultural standards
    company standards
    vendor standards

Display dynamics
- Number of display screens provided is sufficient.
- Sufficient rapidity and ease of access to information.
- Changes easily noticed.
- Easy confirmation that process is operating normally.

Table 9-1 *(continued)*

Display dynamics (cont.)
- Operators can diagnose faults readily by requesting information as needed (avoid alarm flooding).
- Sequence displays are provided for batch and sequential processing.
- Information is displayed at a pace which is adapted to the operator.

Alarms
- Suitable form and location of presentation.
- Number minimized.
- Alarm indications separated from plant status indications.
- Alarms prioritized.
- Alarms grouped.
- Nuisance alarms avoided.
- Return-to-normal indications employed.
- Alarm settings adjusted according to process operating mode.
- Cascade alarms suppressed.
- Choice of presentation modes (panel, VDU).
- Suitable design of audible alarms including auto-silencing.
- Appropriate design of operator alarm acknowledgment.
- Provisions for handling alarm flooding.

Sound
- Used only when it enhances meaning and does not distract.

derstanding user needs and identifying user interactions with the displays using task analysis techniques such as hierarchical task analysis and tabular task analysis for simple tasks and cognitive task analysis for tasks involving diagnosis (Shneiderman and Plaisant, 2003; Mayhew, 1997; Hackos and Redish, 1998). Usability testing is also employed to factor user feedback into the design (Dumas and Redish, 1999; Mayhew, 1999).

The principles and guidelines established in the literature can be used to appraise existing control software designs using human factors checklist methods as described in Chapters 26 and 27. The key issues in this case are the extent to which existing control software designs can and should be changed. Significant changes can disrupt process operations and possibly introduce new problems. Any modifications should be made carefully and with appropriate review and testing.

A checklist of key issues for sensing, understanding and using control software displays is provided in Table 9-1. Figure 9-1 shows poorly designed screens. The top screen shows graphics of nine filters and the bottom screen shows their respective totalizers. The top screen is complex, cluttered, and confusing. Furthermore, the graphics and totalizer screens use different layouts for the filters so there is not a one-to-one correspondence even though the two screens are intended to be used together. Figure 9-2 shows a much improved screen layout.

Figure 9-1: Poorly designed screens. Both screens are poorly organized and the color scheme provides poor contrast. The upper screen shows only the units and not how they are connected.

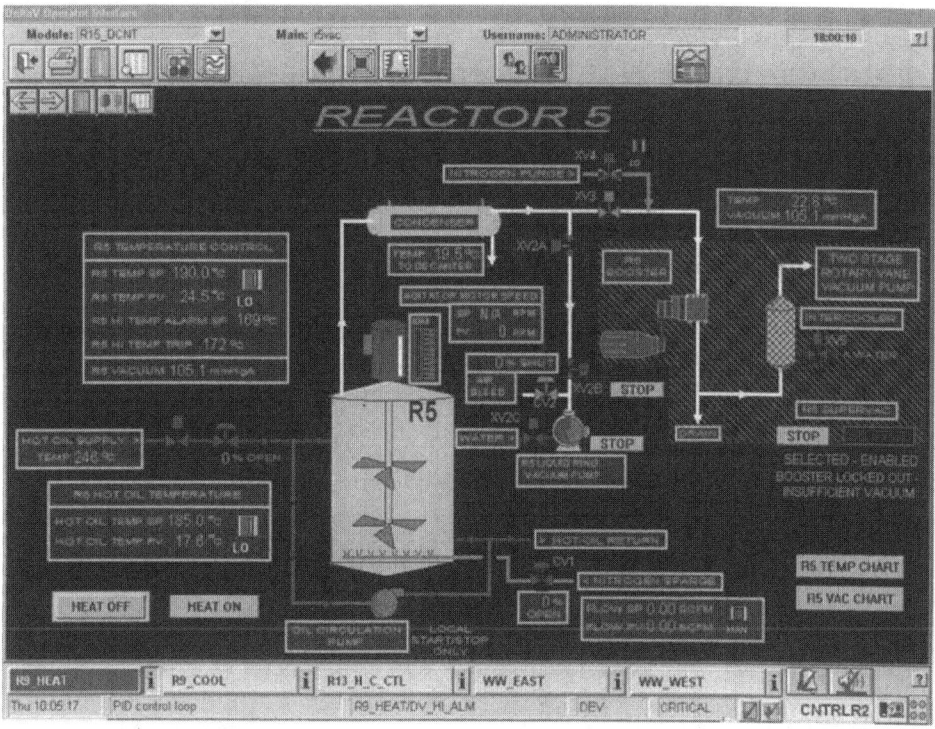

Figure 9-2: An improved operator screen. The screen is well-organized, has good color contrast, and reflects the actual process piping configuration.

9.4 REFERENCES

BS 7179 (1990), "Ergonomics of design and use of visual display terminals (VDTs) in offices" (superceded by BS EN ISO 9241) (London: BSI British Standards).

Def Stan 00-25 (1996), "Human Factors for Designers of Equipment: Part 13: Human Computer Interaction" (Glasgow: UK Ministry of Defence).

Dumas, J. S. and Redish, J. C. (1999), "A Practical Guide to Usability Testing" (Bristol, UK: Intellect, Ltd.).

Gilmore, W. E., Gertman, D. I. and Blackman, H. S. (1989), "The User-Computer Interface in Process Control: A Human Factors Engineering Handbook" (New York: Academic Press).

Hackos, J. T. and Redish, J. C. (1998), "User and Task Analysis for Interface Design" (New York: Wiley).

HSE 60 (1993), Contract Research Report No. 60, "Safety Management of Process Faults: A Position Paper on Human Factors Approaches for the Design of Operator Interfaces to Computer-Based Process Control Systems" (United Kingdom: Health and Safety Executive).

Invergard, T. (1989), "Handbook of Control Room Design and Ergonomics" (New York: Taylor and Francis).

ISA-5.3 (1983), Graphic Symbols for Distributed Control/Shared Display Instrumentation, Logic and Computer Systems (Research Triangle Park, NC: The Instrumentation, Systems and Automation Society).

ISA-5.5 (1985), "Graphic Symbols for Process Displays" (Research Triangle Park, NC: The Instrumentation, Systems and Automation Society).

ISO 9241-1 (1997), "Ergonomic requirements for office work with visual display terminals (VDTs)—Part 1: General introduction" (Geneva, Switzerland: International Organization for Standardization).

Mayhew, D. J. (1997) "Principles and Guidelines in Software User Interface Design" (Upper Saddle River, NJ: Pearson Education).

Mayhew, D. (1999) "The Usability Engineering Lifecycle: A Practitioner's Handbook for User Interface Design" (San Francisco: Morgan Kaufmann).

MIL-STD-1472D (1989), "Human Engineering Design Criteria for Military Systems, Equipment and Facilities" (Washington, DC: U.S. Department of Defense).

NUREG-0700 (2002), Rev. 2, "Human-System Interface Design Review Guidelines" (Washington, DC: U.S. Nuclear Regulatory Commission).

Shneiderman, B. and Plaisant, C. (2003) "Designing the User Interface: Strategies for Effective Human-Computer Interaction," 4th ed. (New York: Addison-Wesley).

9.5 ADDITIONAL REFERENCES

ANSI/IEEE 845 (1988), "Guide to Evaluation of Man—Machine Performance in Nuclear Power Generating Stations, Control Rooms, and Other Peripheries" (Washington, DC: American National Standards Institute).

ANSI/IEEE 1023 (1988), Guide for the Application of Human Factors Engineering to Systems, Equipment and Facilities of Nuclear Power Generating Stations (Washington, DC: American National Standards Institute).

BSR/HFES 100 (2002), Human Factors Engineering of Computer Workstations, Draft Standard for Trial Use (Santa Monica, CA, Human Factors and Ergonomics Society).

Diaper, D. (Ed.) (1989), "Task Analysis for Human-Computer Interaction" (Chichester, UK: Ellis Horwood).

Edwards, E. and Lees, F. P. (1973) "Man and Computer in Process Control" (London, UK: Institution of Chemical Engineers).

Edwards, E. and Lees, F. P., eds. (1974), "The Human Operator in Process Control" (London: Taylor and Francis).

EEMUA (2002), Publication 201, "Process plant control desks utilizing human-computer interface: a guide to design, operational and human interface issues" (London: Engineering Equipment & Materials Users Association).

ISO 9355-1 (1999), "Ergonomic requirements for the design of displays and control actuators—Part 1: Human interactions with displays and control actuators" (Geneva, Switzerland: International Organization for Standardization).

ISO 9355-2 (1999), "Ergonomic requirements for the design of displays and control actuators—Part 2: Displays" (Geneva, Switzerland: International Organization for Standardization).

Laurel, B. (Ed.) (1990), "The Art of Human-Computer Interface Design" (New York: Addison-Wesley).

NUREG/CR-4227, "Human Engineering Guidelines for the Evaluation and Assessment of Visual Display Units" (Washington, DC: U.S. Nuclear Regulatory Commission).

CHAPTER 10

Safe Havens

10.1 INTRODUCTION

Evacuation of plant areas in the event of an emergency such as a fire, explosion or toxic gas release should be addressed in an emergency evacuation procedure—see Chapter 30 on Emergency Preparedness and Response. The procedure should specify designated safe areas, assembly or muster points, shelters, and safe havens (also called safe refuges) for plant personnel. Safe areas are assembly or muster places away from the plant processes where personnel may assemble without being exposed to the immediate danger in order to be accounted for in a roll call. Shelters provide passive protection for their occupants when ventilation is off and windows and openings are closed—most buildings provide some measure of protection. Safe havens are designed to withstand the effects of emergencies such as fires, explosions and toxic releases. They usually double to provide some other function for the facility, e.g. a control room, but they may also be purpose-built. Shelters and havens are usually intended and designed for short-term use and not long-term occupancy. Naturally, since they are designed to protect people, there are numerous human factors issues that should be addressed in their design and use.

In the event of an emergency, plant personnel should follow a prearranged plan or be notified by the plant public address system to evacuate or go to a shelter or safe haven. Sometimes sheltering in place may be preferable to risking exposure while evacuating. The choice depends on various factors, including the location of personnel within the plant (CCPS, 1995).

Plants usually contain a variety of buildings. Some of them must be capable of continued use during an emergency. This includes control rooms, emergency operations centers, emergency command centers, shelters, safe havens, medical facilities, and media information centers. Such buildings may be located remotely from plant processes or they may be hardened to withstand the effects of fires, explosions, and toxic releases. Control rooms are often hardened owing to their frequent proximity to processes. They may be located in separate buildings or be part of another building. Whatever their location, control rooms should be designed to withstand the initial incident in an emergency and allow operators to shut down the process safely or control the process during the emergency. Risks to the occupants of the control room should be within acceptable limits.

Not all buildings may be suitable as safe havens since they may be located too close to hazards or they may not be able to withstand fires, explosions, or toxic releases. Since control rooms are often provided with such protection, they are sometimes designated as safe havens for other plant personnel who may move to them in the event of an emergency.

10.2   HUMAN FACTORS ISSUES

Location

Havens ideally should be located so their exposure to fires, explosions, or toxic releases is minimized while providing proximity to work locations for personnel who need to move to the haven. The prevailing wind direction should be considered when addressing toxic releases.

Signage

Shelters and havens should be marked prominently. Routes may also need to be marked to provide assistance to personnel trying to reach them when seconds may count.

Layout

The size and space within the haven should be sufficient to accommodate the number of people who may use it. An allowance should be made for additional people not anticipated by the emergency response plan. The interior layout of the haven should allow personnel responsible for plant control or emergency response to continue their efforts to control the emergency without interference or distraction from personnel sheltering in the building.

Use

People must know the locations of safe havens, be capable of reaching them in time, and be willing to use them. Protection must not be compromised, e.g. doors propped open for ventilation or convenience, PPE or other emergency equipment/ supplies removed for other uses, escape routes blocked, backup shelters or havens not maintained.

PPE

Sufficient personal protective equipment (PPE) should be available for the personnel working within the haven plus any others who move there during the emergency. In the event the integrity of the building fails, the occupants may need to use the PPE to remain in the building or evacuate to another location. PPE includes SCBAs and bottled air, as appropriate, and it must receive appropriate maintenance.

Protection

Measures to prevent toxic material ingress should be sufficient to avoid the need for personnel to use PPE within the haven, particularly operators in a control room who need to remain unencumbered so that they can better deal with the emergency. Air monitoring equipment should be available to determine the level of hazardous materials present in the atmosphere. Adequate fire and explosion protection should be provided. Emergency lighting and first aid supplies should also be available.

Escape Routes

Routes should be designated in the event the haven must be evacuated.

Backups

Alternative shelters and havens should be designated or alternative procedures specified in the event the primary haven is unavailable or cannot be reached.

Services

Building HVAC, sanitary systems and other utilities should be capable of continued operation during the emergency and be able to provide sufficient capacity for the number of people present. Emergency power supply systems should be provided for process operation and shutdown.

Communications

People inside the haven should be provided with the means to communicate with other plant personnel and emergency personnel located elsewhere so they may stay informed on the progress of the emergency and be provided with further instructions as needed. Control room personnel should be able to provide information about the process to plant management and emergency responders. Communications should be possible while wearing PPE.

Accounting for Personnel

Roll calls should be taken to identify missing persons and communicate their names to emergency responders. Responsibility for these actions should be assigned.

Training

Drills and practices should be conducted to ensure safe haven procedures work and personnel can follow them without difficulty.

> A plant control room was designed as a safe haven. A hazardous liquid released from the plant entered the control room through the sewer system prompting evacuation of the control room.

10.3 TOOLS

Buildings must conform with applicable industry building standards and local codes.

The human factors issues described above can be incorporated into design criteria for safe havens. A number of publications provide guidance on designing and determining the capability of occupied buildings to withstand fires, explosions, and toxic releases (UK Chemical Industries Association, 1998; CCPS, 1996; API RP 752, 1995).

Issues to address for explosions and pressure bursts include the protection of windows, the presence of heavy equipment on roofs (e.g. air conditioners), and the ability of internal fixtures to withstand shaking. Issues to address for toxic materials include adequacy of building seals to prevent toxic material ingress, and the design of the building ventilation system to prevent toxic material intake, possibly by preventing air intake during an emergency. The building sewer system design should also prevent the ingress of hazardous materials into the building. Fire protection measures should ensure the building will withstand thermal radiation effects without collapse and that ingress of smoke and combustion products is controlled. Materials of construction should be fire resistant for the duration of fire events. Smoke and combustion product ingress may be controlled in a similar manner to toxic material ingress.

Hazard analysis and risk analysis methods should be used to assess the risks for safe havens, including human factors considerations. See Chapter 26 on Qualitative Hazard Analysis and Chapter 27 on Quantitative Risk Assessment for more information.

10.4 REFERENCES

API RP 752 (1995), "Management of Hazards Associated with Location of Process Plant Buildings" (Washington, DC: American Petroleum Institute).

CCPS (1995), "Guidelines for Technical Planning for On-Site Emergencies" (NY: AICHE Center for Chemical Process Safety).

CCPS (1996), "Guidelines for Evaluating Process Plant Buildings for External Explosions and Fires" (NY: AIChE Center for Chemical Process Safety).

UK Chemical Industries Association (1998), "Guidance for the Location and Design of Occupied Buildings on Chemical Manufacturing Sites" (London: UK Chemical Industries Association/Chemical Industry Safety, Health and Environment Council).

CHAPTER 11

Labeling

11.1 INTRODUCTION

Labeling is an extremely important aspect of facility design. Labels should be provided whenever personnel are required to identify equipment, follow procedures, or avoid hazards, except when it is obvious to the observer what an item is and how it may be used. Benefits of proper labeling include:

- Reduced learning time for new operators.
- Reduced errors of commission with using the wrong piece of equipment or performing the wrong action.
- Decreased reaction time in abnormal situations.

Labels should be consistent with the following factors:

- Accuracy of the identification/information.
- Time available for label recognition or other responses.
- Distance at which the labels must be read.
- Illuminant level and color.
- Criticality of the function.
- Consistency of label design within and between systems.

The term label means any type of plate, sign, placard, inscription, legend, marking, or combination of these, that is used for purposes of identification, or to impart information or instructions to the reader. There are several types of labels:

- **Hazard label:** Used to identify and provide information about existing or potential situations that may be hazardous to personnel or equipment. For instance, a label might identify a high voltage hazard.
- **Danger label:** A type of hazard label used to identify and provide information about a situation in which an action or omission of action could result in serious injury or death, serious damage to vital equipment, or a major reduction in the facility's capability to perform its mission. For instance, a danger label might identify a non-smoking area due to the presence of flammable solvents.

- **Warning label:** Used to indicate a location, equipment, or system in which a potential hazard exists that is capable of producing injury or death to personnel if approved procedures are not followed. For instance, a warning label might identify an area requiring a confined space entry procedure.

- **Caution label:** Used to identify and provide information about a situation in which an action or omission of action could result in a minor injury, minor damage to equipment, or a minor reduction in the facility's capability to perform its mission. For instance, a caution label might identify the possibility of a wet floor with the potential for a slipping hazard.

- **Identification label:** Used for identification purposes only. These labels identify things such as spaces, locations, equipment, systems, controls and displays, and classifications. For instance, an identification label might identify a process line as one that contains a specific, hazardous chemical.

- **Instruction labels:** Used to present step-by-step instructions for accomplishing a specific task, which is usually operation or maintenance related, and to provide hazard and safety information related to performing the task. For instance, an instruction label might indicate that a door must be kept closed.

- **Information label:** Used to present non-procedural information of a general nature related to health, first aid, sanitation, rules, housekeeping, general conduct, etc. For instance, an information label might identify the location of a safety shower.

- **Graphic label:** Used to present information through line schematics, diagrams, charts, tables, pictures, and so on. For instance, a graphic of a fire extinguisher might be used to identify the location of a fire extinguisher.

A maintenance worker was instructed to repair pump 5 in a series of five pumps. Unfortunately, the pumps were physically layed out in the order 1-2-3-5-4 instead of the logical order of 1-2-3-4-5. The pumps were not labeled. The worker opened pump four instead of pump five and flammable solvent escaped from the pump.

11.2 TOOLS

Each unit, assembly, subassembly, and part should be labeled with a clearly visible, legible, and meaningful name, number, code, mark, or symbol as applicable. The gross identifying label on a unit, assembly, or major subassembly should be located:

- Externally in a position that is not obscured by adjacent items.
- On the flattest, most uncluttered surface available.
- On a main chassis of the equipment.

- In a way to minimize wear or obscurement by grease, grime, or dirt.
- In a way to preclude accidental removal, obstruction, or handling damage.

Labels should be horizontal and read from left to right. When space is limited and the label is not critical for personnel or equipment safety, a vertical label may be used that should read from top to bottom. Labels should not be placed on controls that turn (except handwheels) to prevent the label from being placed in an upside-down position.

Safety and hazard information that is combined with other information (such as instructions) on labels other than hazard labels should be designed such that the safety and hazard information is visually distinct from the other information, and is easily identified by the user.

References directing the user to other documents or sources of information should not be made in a label if the referenced material is required to complete a task, recognize or avoid a hazardous situation, or clearly or completely understand the message of the label. References to other documents or sources of information may be used only to direct the user to supplemental information.

Trade names, company logos, equipment model and/or contract numbers, or any other information not directly required by the user to accomplish his/her operational task(s) should be avoided on the face of a display, control panel, console, or any

Figure 11-1: An example of what can happen if proper labeling guidelines are not followed. The start/stop switches are labeled in reverse, and someone decided that the pump was actually the feed pump rather than the return pump.

other location in direct visual access to the user. Where such information is provided it should be on the side or back of the display or console.

An example of what can happen if these guidelines are not followed is shown in Figure 11-1.

11.3 REFERENCES

American Bureau of Shipping (2003), "Guidance Notes For the Application Of Ergonomics To Marine Systems" (Houston, TX: American Bureau of Shipping).

ASTM F1166-95a (2000), "Standard Practice for Human Engineering Design for Marine Systems, Equipment and Facilities" (West Conshohocken, PA: American Society for Testing and Materials).

MIL-STD-1472F (1999), "Human Engineering" (Washington, DC: U.S. Department of Defense).

People

CHAPTER 12

Training

12.1 INTRODUCTION

Training provides skills and/or knowledge for the current job. Many people must be trained in a variety of areas in order for a process safety management (PSM) program to function properly. This includes operators, mechanics, safety personnel, engineers, supervisors, managers, administrators, contractors, and others. The type of training required includes skills-based, technical, managerial, and awareness. Forms of training include: classroom, using presentations and/or videos; computer-based and web-based; simulators; and on-the-job.

There are many human factors issues that affect training. This includes the design, form, quality, and use of training materials; alignment of training with actual jobs; instructor performance; instructional methods; frequency and content of training; use of training time; training validation; and the training environment. Guidelines that address these issues are provided below for establishing a training program and designing and delivering training.

> A company did an extensive research project and found that the accident rate for the first six months of a new work assignment was six times higher than after the first six months. They also found that this applied to new employees as well as employees changing their job function within the company. Therefore, it is extremely important to provide both new employees and those changing their job functions with a comprehensive training program extending over those first six months, with at least part of that program addressing any important human factor issues that may need reinforcement.

12.2 GUIDELINES FOR TRAINING PROGRAMS

An overall training program should be established as part of an organization's management system. Key guidelines for the program are:

- The program must address regulatory requirements and industry standards.
- Management must believe the training is important and be willing to send personnel to training.

- Training requirements for employees should be established for both initial and on-going refresher training.
- Job descriptions should be correlated with training requirements.
- Performance requirements and standards for trainees should be established.
- A method must be specified for trainees to demonstrate that they have mastery of the training.
- Standards for instructors should be established.
- Training should be designed for the specific functions and jobs to be performed.
- Site-specific training should be used when appropriate.
- Periodic refresher training should be provided, and it should emphasize skills that are not practiced routinely such as diagnosis and response to upsets.
- Refresher training should not be the same as initial training, and it should vary from one delivery to the next.
- The training should incorporate new knowledge and information, changes, and lessons learned.
- Records should be maintained.

Examples of operator, maintenance, and technical staff training programs have been published (CCPS, 1992).

12.3 GUIDELINES FOR DESIGNING AND DELIVERING TRAINING

An example of a typical design cycle for training is provided in Figure 12-1.

Preparation

Pre-training study materials should be provided and attendees should be encouraged to complete the assignment and given the time to do so.

Trainees

Trainees should meet established prerequisites. Trainees should be analyzed to determine appropriate training content based on their education, experience, current skills, and knowledge, previous training and related skills, and attitude/motivation. Training should address diverse learner groups, as appropriate.

Instructors

Trainers should have appropriate training skills and knowledge of the topic.

Logistics

Ideally, training locations should be in the workplace or simulate it.

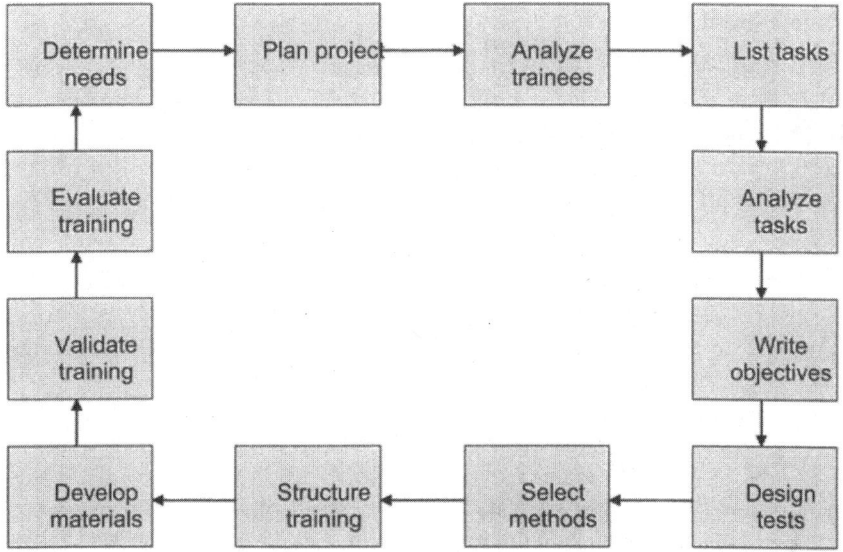

Figure 12-1: Example of a Design Cycle for Training.

Design and Structure

Training should target those meeting the minimum prerequisites. Generally, training should be structured to provide an introduction, content presentation, practice in application by trainees, practice monitoring by trainers, provision of feedback to trainees from trainers, and a summary. The structure of the job should be followed.

Appropriate media should be selected and the duration of training determined. Audio-visual materials and job aids to be used, and job equipment needed should be identified. Training content should be sequenced appropriately, generally going from the known to the unknown, the simple to the complex, and the concrete to the abstract. Pilots should be conducted to refine new training.

Content

The best available subject matter experts who are doers rather than supervisors should be consulted and managers should help determine content. Task analysis and content analysis should be used. Task analysis identifies actions needed. Content analysis identifies knowledge needed to perform the actions. The level of detail provided should target trainees meeting the minimum prerequisites for the training.

Objectives

Actions required of trainees when performing a task should be specified, including the conditions under which the task is performed and the standards that must be met for successful performance.

Instructional Methods

A variety of instructional methods and media should be used. Instructional methods should be selected that allow the trainees to reflect on the content and determine how it applies to their job, see the task being performed, be physically active and perform the task, hear instructions spoken, read materials, take notes, reason through real-life situations, and interact with each other.

Materials

Instructor guides, presentation materials, course manuals, handouts, job aids, and so on. should be prepared. Formats should be varied and the most efficient formats used.

Delivery

The focus should be on tasks that are important and difficult to learn while ensuring the basics are understood. One-third of the training time should be spent on presentation and two-thirds on application and feedback. Reviews and summaries should be provided frequently.

Practice Methods

Tasks should be practiced wherever possible. Exercises should be as realistic and job-like as possible. They should include equipment, materials, procedures, and so on. that will be used on the job, as applicable.

Feedback

Concrete, specific, and objective feedback should always be provided to trainees and given immediately after applications. Positive feedback from the instructor, and other trainees, should be provided whenever possible.

Tests

Methods should be devised to assess whether learning objectives have been met. This may include qualification or certification of trainees to work. There should be a procedure for retraining/retesting and a clear protocol for disqualifying trainees with unacceptable performance.

Developing valid tests and the means to verify understanding of training can be challenging and the people who do so should be suitably qualified.

Validation

The performance of representative trainees on the job should be observed and any performance problems identified. Training should be modified accordingly.

Evaluation and Revision

Trainees should always be requested to provide course evaluations using brief evaluation forms and input should be sought from supervisors on the post-training performance of trainees. Feedback should be used to revise the training, as appropriate.

Environment

The training environment should be chosen to facilitate learning.

Post Training

Trainees should work with an experienced employee who observes performance and provides feedback. Performance reviews should be conducted and the results factored back into the training program.

> Operator training at a company was targeted at an average operator. The company found that some operators had a higher-than-expected error rate in operating the process. They were those in the lower half of the class. The training program was revised to address the least qualified operators.

12.4 TOOLS

Training needs analysis is used to help determine what training is needed (McConnell, 2003). Task analysis (Jonassen, et al., 1998) can help in numerous ways:

- Formulation of training objectives.
- Use in designing tests.
- Help in selecting training methods.
- Assisting in structuring training content and developing materials.
- Helping determine validation and evaluation methods.
- Ensuring nothing is missed.
- Providing feedback to improve procedures.
- Educating management on the complexity of jobs.

Job hazard analysis helps to identify hazards of jobs and any special requirements for performing the job safely (Swartz, 2001). Guidance is available on establishing training programs and designing, delivering, and evaluating training (Goad, 1982; Greber, 1989; Greber, 1990; Rossett and Gautier-Downes, 1991; Stonewall, 1991; CMA, 1993; Wilson, 1994; Kirkpatrick, 1998).

12.5 REFERENCES

CCPS (1992), "Plant Guidelines for Technical Management of Chemical Process Safety" (NY: AICHE Center for Chemical Process Safety).

CMA (1993), "Operations Training Guide for Process Safety" (Washington, DC: Chemical Manufacturers Association, now American Chemistry Council).

Goad, T. W. (1982), "Delivering Effective Training" (San Diego, CA: Pfeiffer and Company).

Greber, B. (1989), "Evaluating Training" (Minneapolis, MN: Lakewood Publications).

Greber, B. (1990), "Designing Training" (Minneapolis, MN: Lakewood Publications).

Jonassen, D. H., Tessmer, M. and Hannum, W. H. (1998), "Task Analysis Methods for Instructional Design" (Mahwah, NJ: Lawrence Erlbaum Associates).

Kirkpatrick, D. L. (1998), "Evaluating Training Programs," 2nd ed. (San Francisco, CA: Berrett-Koehler).

McConnell, J. H. (2003), "How to Identify Your Organization's Training Needs: A Practical Guide to Needs Analysis (NY: American Management Association).

Rossett, A. and Gautier-Downes, J. (1991), "A Handbook of Job Aids" (San Francisco, CA: Jossey-Bass Pfeiffer).

Stonewall, L. (1991), "How to Write Training Materials" (San Diego, CA: Pfeiffer and Company).

Swartz, G. (2001) "Job Hazard Analysis, A Guide to Identifying Risks in the Workplace" (Rockville, MD: Government Institutes).

Wilson, J. B. (1994), "Applying Successful Training Techniques" (Irvine, CA: Richard Chang Associates).

CHAPTER 13

Communications

13.1 INTRODUCTION

Numerous incidents have been attributed to breakdowns in communications—between workers, between work groups, and between managers and workers. For each communication pathway, there must be a commonly understood language with which we can encode/decode the message, a means to convey the message, and, ideally, a means by which errors in the message content can be detected and corrected. This discussion will focus on communications other than written procedures covered in Chapter 22.

13.2 ISSUES/EXAMPLES

The most basic decision in any communication is the language to use. It may be Morse code, semaphore, English, and so on, but both the sending and receiving parties must understand and agree on the language. English is a common choice, but will it be American English, Australian English, British English, Canadian English, Indian English, or South African English? It makes a difference when the worker searches for a flammable gas leak with a "torch."

The issue is further complicated if the plant is located where the chosen language is a second or third language for many of the workers. For example, will written communications be in English while the local language is used for oral communication? Will the procedures be translated into the local language(s) and will all versions be kept up to date? Will equipment labels and warning signs be posted in all relevant languages? It is best if one consistent language is used throughout the plant, but some accommodation must be made for differences in language proficiency. When clear communication is vital to safety, the warning signs and messages should include local terminology.

Once a common language is chosen, then the parties must agree on what specific words mean. For example, if the worker is told to "verify" that a valve is open, does that mean to check its position and open it if she finds it closed, or should she stop and find out why the valve is in the closed position? Either choice is in common usage, but the second choice is much less likely to cause an accident.

The primary tool for insuring a common language is a dictionary, and each plant site or company should develop a customized dictionary of terms and definitions for use in both written and verbal communications. Employees must be trained in the plant lexicon and demonstrate satisfactory mastery of the lexicon. Management must enforce the use of standard plant language and ensure that the dictionary is kept up to date. If management cannot ensure that everyone understands the standard plant language, e.g., on warning signs at a truck loading rack, then standard symbols and color codes should be used to facilitate "universal" understanding.

Given a common language, there must be a defined means to transmit the message from one party to another. Co-workers often use oral communication for routine activities. While that may be fine in the control room, it may be difficult or impossible in the field due to background noise or personal protective equipment such as respirators or ear plugs. When the workers are at separate locations, radios, telephones, or other means must be provided to transmit the verbal information.

Written communications are generally less error prone than verbal communications, and are much preferred for communications between work groups or between management and workers. Two especially critical communications are those between shifts and those between operations and maintenance. For shift-to-shift communication, a written log is commonly passed from one group to the next, and better organizations plan some shift overlap so the outgoing shift can review the shift log with the incoming shift. However, a standard shift change form (as illustrated in Figure 13-1) is far superior to a freehand log or general form with topic headings and freehand notes within. The standard form creates a culture and expectation that important information will be communicated in a standard format. The workers expect to receive that information when they arrive and provide it when they leave their post. Similarly, management should communicate with workers in a standard written format, such as a batch sheet or ticket, so that everyone knows what is being made in which equipment. When changes are made, those should also be communicated to the workers in a standard written format. If those changes are for mainte-

Shift Turnover Log

Date: _5 Nov 2006_   **Shift:** _B to C_

Kettle 101

Current Batch	_369_
Current Step	_15_
Pressure	_150 psi_
Temperature	_312 F_
Cooling water flow	_-_
Steam flow	_43.2%_
Deviations	_None_

Maintenance Work Orders

Completed: _#12347_
_#12402_

Ongoing: _None_

Figure 13-1: An example of a shift turnover log, used to communicate information between shifts.

nance, the written communication must be in accordance with written policies and safe work practices.

Finally, the best communication systems provide some means to detect and correct errors in transmission. For verbal communications, one of the simplest and most effective means is "repeat back." This forces the person receiving the communication to repeat back their understanding to the originator, who can then verify that the other party correctly heard and understood it. If so, the originator authorizes action; if not the originator resends the message and awaits correct confirmation. Written communications are seldom subject to error checking after they are sent; if anything, an acknowledgement of receipt is usually presumed to mean that the message was received and understood.

Because of the high probability of error in communications, management must consider the consequences of error and provide independent protections as necessary to reduce risk to tolerable levels. Physical locks and tags may be used to ensure that miscommunication does not result in a hazardous energy or material release; computer interlocks may be used to ensure that only the correct materials are added to a batch, etc. Procedures, Safe Work Permits and Permit to Work systems are essentially "communication tools"—see Chapters 22 and 24 of this book for additional discussion.

A new pump was added to a plant site. The new pump was given the equipment letters of "JA" as opposed to the older pump lettering of simply "J." A maintenance worker was verbally told to repair pump JA101, but he mistakenly began work on pump J101 and a release occurred. It is difficult to distinguish between the verbal pronouncement of J and JA.

13.3 TOOLS

Field observations and audits are the most common tools for ensuring that appropriate means of communication are available to the workers. When asked, workers usually know where they cannot hear verbal communications, where there are "dead spots" or interference on the radio, and where communication systems are simply overloaded so that timely communication is impossible during a process upset. An "articulation index" (Sanders, 1993) can be used to evaluate communications in noisy environments.

13.4 REFERENCE

Sanders, M. S. and E. J. McCormick (1993), *Human Factors in Engineering and Design,* 7th ed. (New York: McGraw-Hill) pp. 206–208.

CHAPTER 14

Documentation Design and Use

14.1 INTRODUCTION

A wide variety of documentation is used in process safety management (CCPS, 1995). Documents used include policies and procedures, plans, manuals, drawings, diagrams, charts, guidelines, checklists, data sheets, logs, records, work orders, reports, forms such as those used in permits-to-work and management of change, and training materials.

While the types of documents needed for process safety have been thoroughly described (CCPS, 1995), documentation design and use have largely been ignored, other than for procedures (Wieringa, et al., 1998). This is unfortunate since the interactions of people with documents involve human factors issues that can have major impacts on process safety. Policies that are not understood, procedures that are not followed, guidelines that are not used, diagrams that are misleading, records that are not completed properly, forms that are incomplete, and training documentation that is poorly designed can all increase the likelihood of accidents.

Users of documentation bring with them many attributes that should be considered when designing and developing documentation, including skills and knowledge, abilities, habits, mindsets, experiences, motivations, attitudes, culture and social conditioning. These attributes result in needs and expectations that must be taken into account when developing documentation. Also of importance are how humans sense and perceive information, access information by reading, learn, remember (and forget), solve problems, and act. Often, personnel in an organization will share common conceptions, called schemas or schematas. However, variations in personnel attributes and schemas should also be addressed.

Key issues for document design are:

- Medium used in the document.
- Navigation in the document.
- Presentation of information.
- Content of information.

Users of documentation must interact with the medium and navigate through the presentation to reach the content before they can assimilate the information (see Figure 14-1). The medium and methods of navigation and presentation used can help or hinder users in reaching and assimilating the content.

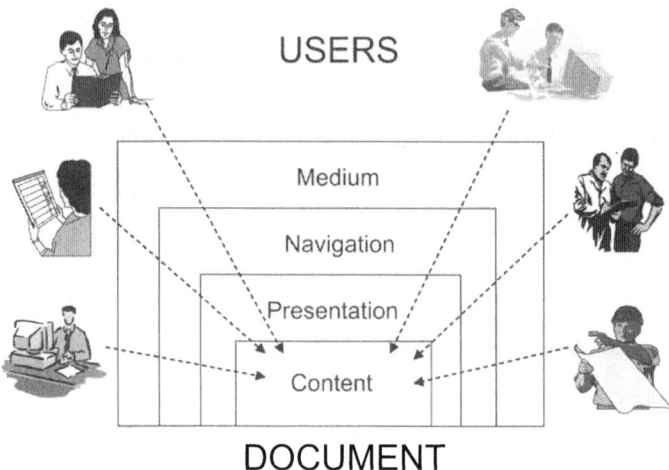

Navigation provides the paths by which users move around the information in the documentation. If users spend less time accessing information, they have more time to assimilate it. Also, if the information is difficult to locate, the user becomes frustrated, may give up trying, and may resort to assumptions that may be wrong.

Presentation is how information looks on a page or screen. It includes layout, which is the physical arrangement of text and graphics, the use of color, and the fonts employed. Presentation should provide perceptual clues that enable users to process information. Users interpret presentations before they begin the cognitive tasks of reading, remembering, and understanding the content of the information, that is, the text of the information presented.

The first page of a well-designed operating procedure is shown in Figure 14-2.

14.2 CONVERTING TO ELECTRONIC DOCUMENTATION

There is a trend toward increased use of electronic media for documentation. For example, companies are using electronic operating procedures and converting legacy information into on-line manuals and help systems. Hard copy has distinct advantages, including schemas for navigation and presentation that are almost universally understood, portable, and easily annotated. However, electronic documentation is less expensive to produce, distribute, and maintain. It is also easy to update, easily found, instantly accessible, and offers efficient navigation techniques. Electronic documentation requires computer hardware and software, more sophisticated users, and does not lend itself to portability. Most significantly, standards for the infrastructure of electronic documents are still evolving.

CCPS Chemicals Inc.
Batch Processing Division

OPS-NO-001 Rev. 0
Operating Procedure

CHARGING METHYL ETHYL KETONE (MEK) TO REACTOR (R-2) AND DRUMMING FINAL PRODUCT

APPROVED BY: ______________________ DATE : ______________
Manager

Table of Contents

PURPOSE

This procedure describes how to charge Methyl Ethyl Ketone (MEK) from a 55-gallon drum into Reactor (R-2) using vacuum. After additional custom batch instructions have been completed, this procedure describes how to transfer processed material from (R-2) into 55-gallon drums. (R-2) is a 100-gallon, jacketed, glass-lined reactor pressure rated at 25 psig at 650°F. Reactor safety features include:

- A sight glass
- 2-inch safety valve set at 25 psig
- 2-inch 316 stainless steel rupture disc set at 22 psig

Date Printed: 1/2/2003 OPS-NO-001 Rev. 0 Page 1 of 13

Figure 14-2: The first page of a well designed operating procedure. The entire procedure can be found in D. A. Crowl, *Understanding Explosions* (AICHE/CCPS, 2003).

Key issues in moving from hard copy to electronic documentation include:

- What makes most sense from the user's perspective?
- What do users want, e.g. level of detail?
- Where and how are users going to use the information, e.g. field versus control room, performing actions versus referencing?
- What type of information is being presented?
- How often does the information need to be updated?
- How often do users access the information?
- Have users been trained in the use of electronic documentation?
- How easy is it for users to use electronic documentation, e.g. forwards and backwards referencing?
- Do users need to skim, scan, search, read for content, or read for evaluation?
- Do users want access to information randomly or sequentially?
- Can users generate hard copy of electronic information when needed?

Hard copy information must be restructured and possibly re-written to convert it for electronic use. Developers should be trained in creating the electronic document and users must know how to use it.

14.3 USE OF DOCUMENTS

It is important that documentation be kept up-to-date and a document management system should be used to help ensure this happens. Similar documents should use consistent formats. The form of documents is also important because it impacts on how effectively they can be used. For example, if work must be carried out in difficult or hostile environments, hard-copy media may need to be protected using plastic lamination. Lengthy documents are unlikely to be of use in the field. Simple checklists will be of greater value.

In some cases, controlled documents are used to ensure that the correct version is used and possibly to limit access to authorized users. Controlled documents should be used judiciously since they create various problems such as the need to ensure all copies are updated when changes are made. The designation of electronic versions as the controlled versions can help overcome these problems.

A detailed operations procedure was developed for a batch process. The resulting document was highly detailed, almost a hundred pages long, and was structured serially so that the batch began on the first page and ended on the last. It was soon observed that the operators, particularly toward the end of the batch, were not following the procedure.

Table 14-1: Checklist for the Design of Documentation

Medium
- Match the medium to the information.
- Consider user needs.

Navigation
- Use navigation clues such as page numbers, running headers and footers, tables of contents, indexes, references and cross-references.
- Use navigation clues for electronic displays such as hyperlinks and interface controls such as buttons and check boxes.
- Use navigation clues that are clear, recognizable, consistent and meaningful.
- Use no more than three layers of information.
- Avoid circular references.

Presentation
- Provide document identification and control information including document title, revision number, and issue date.
- Use a suitable reading level for intended users.
- Use the primary language of the users.
- Use appropriate terminology.
- Define special terms.
- Avoid abbreviations and acronyms.
- Define abbreviations and acronyms where used.
- Use suitable line and sentence lengths, line spacing, letter kerning and margins.
- Use an appropriate type size, typeface, font, and case accounting for the vision of users and lighting conditions.
- Number items in a list only if they are sequential actions or instructions.
- Use bullets or dashes for items that can be performed in any sequence.
- Number pages in the format "page 1 of 10".
- Use graphics, icons, fonts, colors, line weights and emphasis to provide visual cues. Note: emphasis includes bolding, italicizing, underlining, capitalization, initial capitalization, and type size.
- Use visual cues judiciously to avoid sensory adaptation.
- Use check-offs to confirm actions have been taken.
- Use open page layouts. Avoid dense text.
- Use white space proactively to indicate organization, separate modules of information, convey relationships, emphasize pieces of information, direct users' focus, and afford an open look and feel to the information.
- Target 20 to 40% white space for hard copy and 40 to 60% white space for electronic documents.
- Group information together to make it stand out and easier to remember.
- Incorporate lists and other ways of grouping information. Keep lists to 5—9 items.
- Use clear, well-defined hierarchies to communicate relationships in information.
- Use distinct, consistent headings set apart from the body of text and distinguished by font, type size, or emphasis.
- Separate different types of information.
- Ensure users know the type of information being communicated. Use explicit labeling or implicit structuring.
- Use labeled graphics to support or replace text. *(continued)*

Table 14-1: Checklist for the Design of Documentation *(continued)*

Presentation (cont.)

- Do not add graphics for their own sake.
- Position graphics in the user's left visual field.
- Use tables, as appropriate.
- Use symmetrical layouts.
- Ensure user needs and expectations guide the use of color.
- Generally avoid using color in text.
- Use color in graphics but only to enhance content.
- Minimize the number of colors used.
- Use adequate contrast among colors used and between the background and foreground objects.
- Pair colors appropriately, e.g. black on white, blue or red on white, white on black, yellow or green on black.
- Assign colors meaningfully accounting for their cultural, social and emotional associations.
- Never use color as the only coding.
- Always test color under the range of possible viewing conditions.
- Account for color-vision deficiency:
  - Always use color redundantly.
  - Do not use red-green and blue-yellow combinations.
  - Use colors with different lightness values.
  - Limit the number of colors used.
  - Put color legends close to the colors they decode.
- Use a delimiter to identify the end of the document.

Content

- Identify clearly what users must do.
- Provide sufficient information.
- Provide accurate information.
- Provide only information really needed by the user.
- Provide assistance to users from a problem-solving rather than a functional perspective.
- Anticipate and overcome users's obstacles to problem solving.
- Relegate supporting details to appendices.

14.4 TOOLS

Documentation should be designed from the perspective of users, taking into account how the users sense, perceive, learn, remember, access information, read, solve problems, and act. Information should be presented in ways that match users' experiences, expectations, assumptions, knowledge, and physical, mental and psychological characteristics, to help them assimilate the information. Documentation developers must understand document users, their needs and expectations, and work with them during the process of developing documentation. This should culminate in usability testing (Rubin, 1994; Dumas, et al., 1999). Documents should be tailored to their users wherever possible. If only one version of a document is produced, it should be tailored to the lowest common denominator users.

Table 14-2: Checklist for Electronic Documentation

- Provide users with the most direct access to information that is possible. Do not require more than three layers to access information.
- Use more white space and grouping of information than for hard copy information.
- Use lists, tables and graphics to gain more white space and grouping of information.
- Use text and graphic hyperlinks to improve access to information.
- Avoid overuse of hyperlinks. Typically use up to four hyperlinks per 25 lines of text.
- Provide a history feature for access to recently-displayed panels.
- Provide for key-word searches. Anticipate key words users will employ in searches and provide an index.
- Provide a synonym file for key words.
- Provide interactive tutorials.
- Use color but no more than six colors on one panel, including the colors of the background, text and hyperlinks.
- Be consistent in the use of color coding from panel to panel.
- Design to the smallest screen size employed by users.
- Ensure users do not have to scroll to see all of a table or graphic.
- When using screen captures as figures, make it obvious they are not the real thing since users may try to interface with them as though they were. Use borders, size, or color to distinguish them.

Information on documentation design can be found in a variety of sources in disciplines such as human factors (Coe, 1996), technical communications (Schriver, 1996; Hackos, 1994), graphic design (Berryman, 1990; Tufte, 1983), instructional design (Stonewall, 1991), and software user interface design (Mayhew, 1999; Hackos and Stevens, 1997). There is no one source that provides guidance applicable to process safety documentation. Consequently, a set of guidelines has been prepared and is provided in Table 14-1. Specific guidelines for electronic documents are provided in Table 14-2. These guidelines can be used either in preparing new documents or in reviewing existing documents to identify possible improvements.

In today's world many companies operate internationally, and it is important to recognize and account for cultural differences in communications using documentation (Hoft, 1995).

14.5 REFERENCES

Berryman, G. (1990), "Notes on Graphic Design and Visual Communication" (Menlo Park, CA: Crisp Publications).

CCPS (1995), "Guidelines for Process Safety Documentation" (NY: AICHE Center for Chemical Process Safety).

Coe, M. (1996), "Human Factors for Technical Communicators" (Hoboken, NJ: Wiley).

Dumas J. S. and Redish, J. C. (1999), "A Practical Guide to Usability Testing" (Bristol, UK: Intellect Ltd.).

Hackos, J. T. (1994), "Managing Your Documentation Projects" (Hoboken, NJ: Wiley).

Hackos, J. T. and Stevens, D. M. (1997), "Standards for Online Communication" (Hoboken, NJ: Wiley).

Hoft, N. L. (1995), "International Technical Communication" (Hoboken, NJ: Wiley).

Mayhew, D. J. (1999), "The Usability Engineering Lifecycle: A Practitioner's Handbook for User Interface Design" (Oxford, UK: Morgan Kaufmann, now Elsevier).

Rubin, J. (1994), "Handbook of Usability Testing: How to Plan, Design, and Conduct Effective Tests" (Hoboken, NJ: Wiley).

Schriver, K. A. (1996), "Dynamics in Document Design: Creating Text for Readers" (Hoboken, NJ: Wiley).

Stonewall, L. (1991), "How to Write Training Materials" (San Diego, CA: Pfeiffer and Company).

Tufte, E. R. (1983), "The Visual Display of Quantitative Information" (Cheshire, CT: Graphics Press).

Wieringa, D., Moore, C., and Barnes, V. (1998), "Procedure Writing: Principles and Practices," 2nd ed. (Columbus, OH: Battelle Press).

14.6 ADDITIONAL REFERENCES

Hackos, J. T. and Redish, J. C. (1998), "User and Task Analysis for Interface Design" (Hoboken, NJ: Wiley).

Marcus, A. (1991), "Graphic Design for Electronic Documents and User Interfaces" (NY: Addison-Wesley).

Mullet, K. and Sano, D. (1994), "Designing Visual Interfaces: Communication Oriented Techniques" (Upper Saddle River, NJ: Prentice Hall).

Nielsen, J. (1990), "Designing User Interfaces for International Use" (Amsterdam, Holland: North-Holland Publishers).

CHAPTER 15

Environmental Factors

15.1 INTRODUCTION

There are many factors in the process environment that affect the comfort, performance and health of process workers. Four stand out as most important—lighting, noise, temperature and vibration. The following sections provide the comfort ranges for each of these factors and discusses the adverse effects on health, performance and comfort due to working in out-of-limits conditions.

15.2 NOISE

Excessive noise can have several adverse effects, with the most severe being hearing damage. Noise can also interfere with speech, affect performance in tasks requiring concentration, mask warning signals, and cause considerable annoyance and stress. The effects of typical levels of noise on the communications environment for offices, conference rooms, and plant workplaces, adapted from work by Baranek (1989), are listed in Table 15-1.

The perceived intensity of noise is determined by the sound pressure level. This is usually measured in decibels (dB). Table 15-2 illustrates the sound pressure level of common noises.

The human ear is sensitive and delicate. Prolonged exposure to noise over 80 dBA or repeated impulses from, for example, hammering or riveting, can result in permanent hearing loss. Figure 15-1 shows typical noise induced loss of hearing. It shows that under normal exposure, older humans lose the ability to hear high frequency sounds. This has an implication on the design of auditory displays for aging process workers. Well-established noise exposure limits have been developed for both continuous and impulse noise to reduce the probability of hearing damage (Baranak, 1989).

Noise control options include, in order of preference:

- Reducing the noise at its source.
- Wearing hearing protection.
- Limiting exposure time by job rotation.

Different sound frequencies have different effects on people. Noise assessment should always take into account the frequency range of the noise involved.

Table 15-1: Effects of noise on performance

SPL (dBA) Sound Power Level	NC Noise Criterion	Effects on Telephone Use	Effects on Face-to-face communication	
			Situation	Distance
40	NC30	None	Quiet Office Conference table 15 feet	Normal voice range 10–30 feet
44	NC35	None	Conference at table 6–8 feet	Normal voice range 6–12 feet
50	NC40	None	Conference at table 4–5 feet	Raised voice range 6–12 feet
53	NC45	None		
58	NC50	Satisfactory	Conference of 2–3 people	Raised voice range of 3–6 feet
62	NC55	Satisfactory		
67	NC60	Slightly difficult	Very noisy office	Raised voice range 1–2 feet
72	NC65	Difficult		
77	NC 70	Very Difficult	Communication is extremely difficult	

15.3 VIBRATION

The human body can be considered a dynamic mechanical system. When the body is subjected to vibration forces, the vibration energy is absorbed by the tissues and organs. Under extreme conditions, organs can be damaged. Excessive mechanically-induced vibrations can cause muscle fatigue, reduce motor and visual performance, and can cause minor annoyance and stress.

The adverse effect of vibration varies with the exposure duration, frequency, amplitude, acceleration, direction, and point of application. Vibration is measured with accelerometers oriented in each of the three orthogonal axes (tri-axial accelerometers). Figure 15-2 shows a tri-axial accelerometer mounted in a seat pad that measures the vibration transmitted to drivers at the seat pan. Data from the transducers can be compared with accepted standards to determine if the vibration can affect health or safety (ISO, 1997).

Whole body vibration is associated with tasks such as driving. Most whole body vibration is low frequency and affects the body differently at varying amplitudes and direction of application. For example, low frequency vibration (1-2 Hz) applied vertically on the seated body can load the spine and cause back pain.

Segmental vibration is associated with use of power tools (Figure 15-3). This vibration is typically high frequency with low amplitudes. Potential effects associated with repeated segmental vibration are injury to blood vessels, nerves, bones, joints, and muscles. Risk factors associated with segmented vibration are listed in Table 15-3.

Table 15-2: Peak noise level in dB(A)

Source	Noise level in dB(A)
Rifle-shot; motor test bench	130
Pneumatic bore-hammer	120
Release of a relief device	115–120
Rocking sieve; chain saw; compressed air riveter; electric cutter; compressed air hammer	105–115
Fin fans	100
Control valves (steady-state process)	>100
Furnace air (naturally induced)	90
Tool making machine (running light)	80
Typewriter	65–75
Average loud office	60
Busy office	45–60
Quiet conversation	30

Vibration can be controlled through dampening. For example, seats and suspensions are used to damp out vehicle induced vibration.

Hand tool vibration can be dampened within the hand tool or by wrapping the handgrip with dampening material or by using heavy gloves (Sanders and McCormick, 1993). The effects of vibration can also be controlled by decreasing the time that the individual is exposed to the vibration.

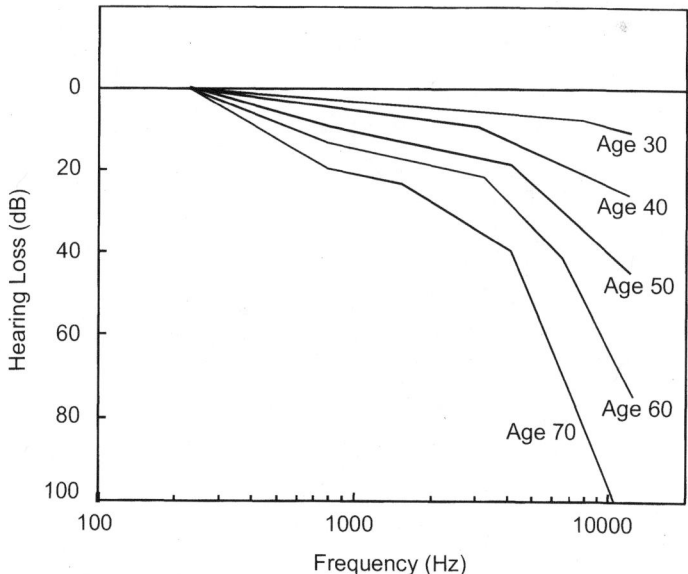

Figure 15-1: Distribution of hearing loss by age

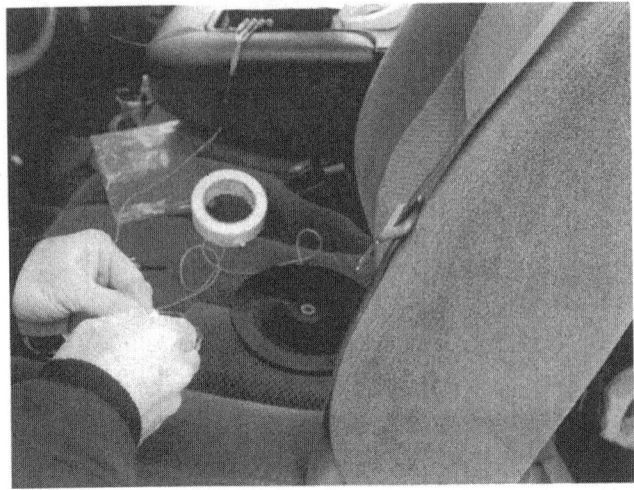

Figure 15-2: Seat pad designed to measure vibration at the seat pan.

Figure 15-3: Use of power tools can cause segmental vibration.

Table 15-3: Risk factors associated with segmental vibration

Physical:
   Dominant vibration frequencies entering the hand
   Years of employment involving vibration exposure
   Total duration of exposure each work day
   Temperature during exposure each day
   Dominant vibration direction relative to hand
Biodynamic:
   Hand grip forces
   Surface area, location, and mass of parts of the hands in contact with the source of
      vibration
   Posture (position of hand relative to body)
   Other factors influencing the coupling of vibration into the hand (e.g., texture of the
      handle)
Individual:
   Factors influencing source intensity and exposure duration:
      tool maintenance,
      work rate, skill and productivity
   Hand size and weight
Prior injury to the fingers or hands

15.4 TEMPERATURE AND RELATIVE HUMIDITY

The human body's thermal regulation system tries to maintain a relatively stable internal (core) temperature of between 97 and 99 °F (36.1 and 37.2 °C). The core temperature must stay within a narrow range to prevent serious damage to health and performance (Attwood et al, 2003). The body maintains the heat balance by increasing or decreasing blood circulation to the skin. The body also exchanges heat with the environment through:

- Convection: absorbing or losing heat from or to the surroundings through the skin.
- Conduction: by contacting sources of heat or cold directly.
- Evaporation: by losing heat through the evaporation of water vapor on the skin.
- Radiation: receiving radiation from an external source or radiating heat from the body.

Clearly, some methods are more effective than others.

When we are hot, our performance suffers. Research indicates that physical activities performed in hot, humid conditions may cause fatigue and exhaustion. Increases in body temperature can also cause an increase in errors in cognitive tasks. Figure 22-4 shows that error rates increase in a wide variety of cognitive and psychomotor tasks as the working temperature increases beyond 98 degrees F (36.6 de-

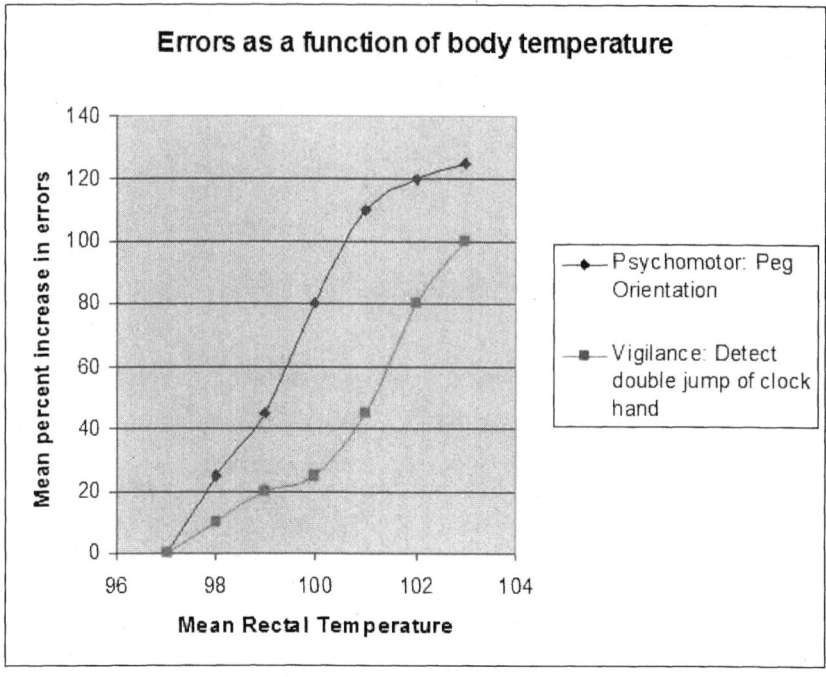

Figure 15-4: Errors rates in workers as a function of temperature for various types of tasks. Figure adapted from Poulton (1972).

grees C). Achieving comfort and performance in hot environments requires changes to the task, personal protective equipment and the individual. Recommendations for working in hot environments include:

- Workers new to a hot environment need to be acclimatized by initially spending only 50% of their time in the heat and increasing this by 10% each day. The American Conference of Governmental Industrial Hygenists (ACGIH) have published guidelines that specify work/rest schedules as a function of ambient temperature and work load to minimize the physiological effects of heat stress on both acclimatized and un-acclimatized workers.
- Workers should be trained to recognize the signs of heat stress.
- The higher the ambient temperature, the less physical effort can be expended.
- Protect the worker from radiant heat with covers over the workplace.
- Increase the flow of air in the workplace using fixed or portable fans.
- Provide cooling vests in extreme heat.

Cold environments can also affect physical and mental performance. Reductions in limb or whole body temperatures affect motor ability, cause a loss of touch sensitivity, affect the limbs' muscular control and reduce dexterity and muscle strength. Gloves and heavy thermal boots and clothing can protect operators in cold weather. In severely cold temperatures, heavy clothes and gloves can also affect the ability of workers to perform tasks that require strength and dexterity by limiting body motion. Recommendations for working in cold environments include:

- Create a protective environment out of the wind.
- Provide portable heaters where possible.
- Provide clothing that is designed for the temperatures experienced.
- Shorten work periods as required by the temperature and wind conditions.
- Provide warming facilities away from the workplace.
- Train workers to recognize the signs of cold stress.

For more detailed guidance on working in hot and cold environments, the reader is directed to Rodahl (1989), Eastman Kodak (1983) and ACGIH (2001).

Effective Temperature (ET) is that combination of temperature and humidity that will produce the same sensation of warmth in individuals (Grandjean, 1988). Research has shown that within the range of 18–24 degrees C (64–75°F) , fluctuations in relative humidity (RH) between 30% and 70% have little influence on ET. When buildings are heated, the relative humidity typically decreases. A RH below 30% can cause irritation of the nasal and bronchial passages. A RH above 70% can produce a feeling of "stuffiness." High prolonged relative humidity can also cause damage to buildings and fixtures.

Many manual material handling tasks require a high level of energy expenditure that can cause physiological stress. Some warehouse tasks, for example, can raise stress under temperate working conditions to the limit of human capability. When the working temperatures rise, the levels of physiological stress can exceed maximum allowable levels requiring increased rest time and reducing productivitiy.

15.5 AIR QUALITY

Air quality can have both long- and short-term effects on people. Low levels of contaminants that can be found in discharges from petrochemical plants can affect the long term health of those exposed. High levels of contaminants can give rise to feelings of distaste and unpleasantness.

Sources of air contamination in indoor environments are odors, dust, fumes, smoke, mists, fogs, vapors and gases that are produced by human occupants, office furnishings and equipment and sources in the outside or replacement air. The quality of the indoor environment can be affected by how well the air conditioning equipment can remove the airborne contaminants and how well the sources of contamination can be identified and reduced. Most buildings today are equipped with systems that filter, wash and disinfect the indoor air. When the systems are working properly, the particulate content of indoor air is likely lower than the "fresh" air from the outside.

Ventilation rates are the flow rates of outside or make-up air that enters the indoor space (Attwood, 1996). The American Society of Heating, Refrigeration and Air Conditioning Engineers (ASHRAE, 1989) sets standards for minimum ventilation rates and air quality that is acceptable to human occupation and are intended to avoid adverse health effects. Minimum ventilation rates for different inside areas have been calculated by ASHRAE and are presented in Table 15-4. The rates specify the minimum percentage of outdoor air that must be mixed with building air.

15.6 LIGHTING

Two characteristics of lighting affect the comfort, performance, and health of the process workers: quantity and quality. The quantity of light required to perform work depends on the characteristics of the visual task and the age of the worker. Light levels are usually expressed as units of illumination which are defined as the amount of light (luminous flux) falling on a work surface. Illuminance is measured in lux (international units) and foot-candles (U.S. units). Figure 15-5 specifies the

Table 15-4: Outdoor Air Requirements for ventilation

Application	Estimated Maximum Occupancy	Outdoor air flow rates cfm/Person (Liters/Sec./Person)
A. Office		
Office space	7	20 (10)
Meeting rooms	50	20 (10)
Auditoriums	150	5 (8)
Cafeterias	100	20 (10)
Reception areas	60	15 (8)
Smoking	70	60 (30)
Lounges		
Restrooms (cfm/wc or urinal)		
Elevators		50 (25)
Lobbies		100 cfm/ft^2
Telecom centers and data entry areas	30	15 (8)
Duplicating	60	20 (10)
		0.5 cfm/ft^2
B. Miscellaneous		
Bars, cocktail lounges	100	30 (15)
Dining rooms	70	20 (10)
School rooms	50	15 (8)
Laboratories	30	30 (10)

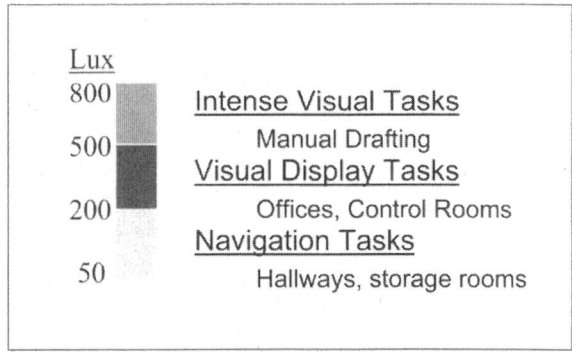

Figure 15-5: Illumination required for various tasks.

illumination required for different tasks. We see the reflection of the illumination from the surfaces that the light illuminates. This is known as luminance or brightness. The brightness of an object depends on the light falling on it and its reflectance. For example, since more light is reflected from a white surface than from a black one under the same level of illumination, the white surface appears brighter than the black one. As a general rule, the more visually intense the task, the more illumination required to perform it.

The quality of light can be expressed in several ways. Glare, for example, occurs whenever one part of the visual field is brighter than the level to which the eye has become adapted. Glare can cause a decrease in visual performance, annoyance, and discomfort. Glare can be direct or indirect as shown in Figure 15-6. Direct glare is caused when a bright light source is in the field of view (Attwood, 1996). A light

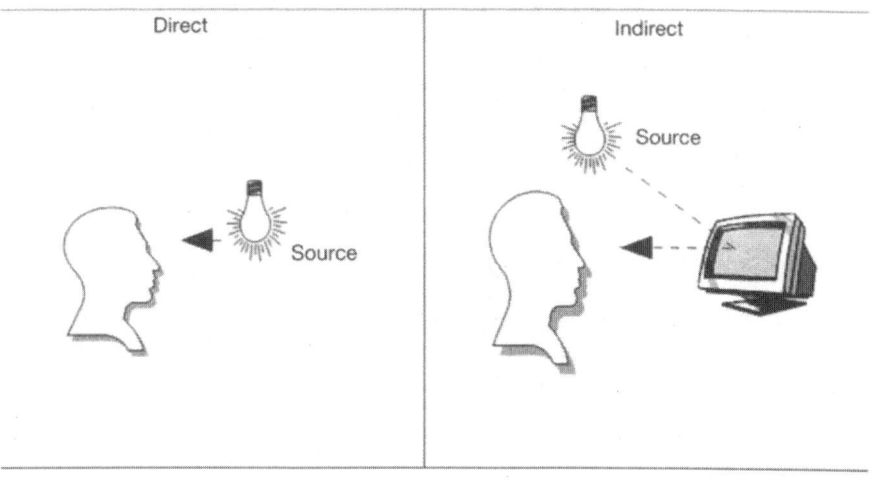

Figure 15-6: Glare caused by indirect lighting.

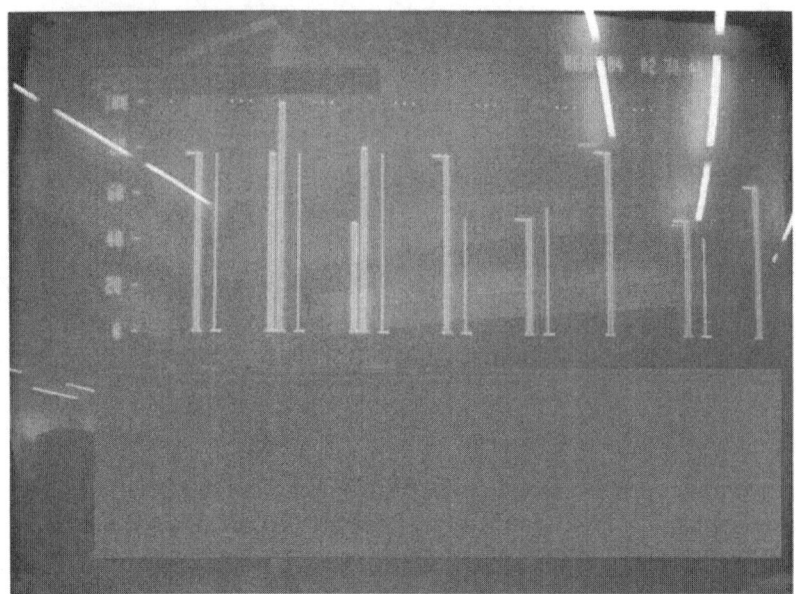

Figure 15-7: An example of a control room screen with glare.

Figure 15-8: Another example of control room glare caused by indirect lighting.

Table **15-5:** Reducing and Controlling Glare

Reduce/Control Direct Glare	Reduce/Control Indirect Glare
Light source should not be directly in the field of vision or line of sight	Position the light source or work area so reflected light is not in the direction in which the person normally needs to look to achieve maximum visibility
Use shades, hoods, glare shields to reduce the brightness (luminance) of the light	Avoid using reflective colors and materials, such as bright metal, glass, glossy paper
Use several low-intensity lamps in place of few high-intensity lamps	Use surfaces that diffuse light, such as textured finishes, non-glossy surfaces, flat paint
When using fluorescent tubes, they should be at right angles to the line of sight	Provide several low-intensity lamps in place of few high-intensity lamps and, use indirect lighting
Use indirect lighting	The brightness of the lamps (luminaires) should be kept as low as feasible

source, such as the sun, located beside an instrument can make the instrument hard to read. Indirect glare, which is also referred to as reflected or specular glare, is caused by a reflection of a bright light source from a polished or glossy surface. Overhead lights reflecting on a process control screen (Figure 15-7) or on the glass face of an instrument (Figure 15-8) will make the information hard to read. Glare can be controlled by redesigning the visual task and the lighting system. Low ambient lighting in controls rooms with computer displays can increase operator comfort and reduce screen glare and eye strain. Table 15-5 (Attwood et al., 2003) provides recommendations for each.

15.7 REFERENCES

ACGIH (2002) "Two thousand and two TLVs and BEIs" (Cincinnati, OH: American Conference of Governmental Industrial Hygienists).

ACGIH (2001), "Heat Stress and Strain: TLV Physical Agents," 7th ed. (Cincinnati, OH: American Conference of Governmental Industrial Hygienists).

ASHRAE (1989), "Ventilation for Acceptable Indoor Air Quality" (Atlanta, GA: American Society of Heating Refrigerating and Air-Conditioning Engineers, Inc.), Standard 62-1989.

Attwood, D. A. (1996), "Office Relocation Sourcebook" (NY: John Wiley and Sons).

Attwood, D., Deeb, J. M., and Danz-Reece, M. E. (2003) "Ergonomic Solutions for the Process Industries [Burlington, MA: Gulf Professional Publishing (Elsevier)].

Baranek, L. L. (1989) "Balanced Noise-Criterion (NCB) Curves." *Journal of the Acoustical Society of America,* 86(2), August, pp 650-664.

Grandjean, E. (1988) "Fitting the task to the man" (London, Taylor and Francis).

ISO (1997), "Mechanical Vibration and Shock—Evaluation of Human Exposure to Whole-body Vibration," Standard 2631-1 (Geneva, Switzerland: International Organization for Standards).

Poulton, E. C. (1972), "Environment and Human Efficiency" (Springfield, IL: Charles C. Thomas, Pubs.).

Rodahl, K. (1989) "The Physiology of Work" (London: Taylor and Francis).

Sanders, M. S. and McCormick, E. J. (1993) "Human Factors in Engineering and Design." (NY: McGraw-Hill).

CHAPTER 16

Workloads and Staffing Levels

16.1 INTRODUCTION

Questions about workload are among the most common and controversial issues in human factors because of their relationship to staffing levels. When asked, employees will almost always claim that they are overworked, and management will almost always claim that there is too much unproductive, idle time. Both views have an element of truth. When a plant is running smoothly, there is often free time that workers can devote to lower priority activities. But during startups, shutdowns, or upsets, the workload and overtime demands may be very high. The "right" staffing level provides adequate resources to do the necessary tasks, and those tasks are distributed in a way that keeps each worker near an optimum stress level.

16.2 ISSUES/EXAMPLES

An individual's physical and mental workload directly affects his/her stress level. Unquestionably, excessive workload leads to human error, because the worker is overwhelmed by the activity. In high-stress situations, the worker may simply cease to pay attention to new data (e.g., alarms) and focus attention on responding to what is known. The worker will skip tasks that he or she judges to be less critical and will rely on skill or rule-based behavior instead of knowledge-based behaviors. This is done in an attempt to reduce the worker's physical or mental workload.

Over the longer term, excessive workload will contribute to a myriad of human errors. Fatigue and inattention to detail are typically observed, and shortcuts become the norm. Cross-checks between workers and work groups are minimized or eliminated, which defeats the protections written into many management systems. Training needs go unmet because there is "no time" and minor incidents are not investigated.

At the opposite end of the scale, very low workloads also lead to human error. If, for example, the computer is performing the work, the operator will mentally disengage from the process and be less effective in responding to process demands or upsets. Bored operators may deliberately push the process beyond stable limits just to challenge themselves, and management may even reinforce this behavior by rewarding workers who set new production records, regardless of how they were achieved.

In summary, the ideal workload must be challenging enough to maintain the worker's attention and interest without overloading the worker.

Another key issue in determining appropriate workloads is the demand of emergency response. Many companies state that they do not staff for emergencies, meaning that they accept the occasional unit trip or shutdown if necessary. However, workers feel immense pressure to recover from upsets and maintain production without tripping the unit or dumping the batch. Even if management does not overtly demand such actions, the workers know that restarting a tripped unit will entail a lot of extra work—they will do everything in their power to avoid it. Management must establish clear requirements to shut down processes when they exceed safe operating limits and these requirements must be enforced.

> In a multipurpose batch facility, critical feed lines required double isolation to avoid cross contamination. To reduce their workload, operators began closing only one valve. Ultimately, one of these valves leaked and the contamination catalyzed a runaway reaction.

16.3 TOOLS

A classic approach to determine appropriate physical workloads is through time-and-motion studies. Workers are observed performing a given task, average times are calculated, and norms/goals are established. Particularly in manufacturing situations, the worker's compensation may be directly tied to achieving the target productivity of X widgets per hour. A variation of this, called activity analysis, is useful for determining the proportion of time spent in different activities. Link analyses diagram the worker motions and determine whether one or more people are required to perform the necessary tasks.

Several tools are available to gauge mental workload, such as:

NASA TLX—NASA Task Load Index (Hart and Staveland, 1988)

OWL—Operator Workload (subset of TLX) (Hill, et al., 1987)

PROCRU—Procedure Oriented Crew Model

SWAT—Subjective Workload Assessment Task (Reid and Nygren, 1988).

SWORD—Subjective Workload Dominance (Vidulich, et al., 1991)

TAWL—Task Analysis/Workload

TLAP—Time-Line Analysis and Prediction

W/INDEX—Workload Index

In chemical facilities there are many different tasks that must be performed, and the circumstances are varied—different approaches have been developed to gauge workload. Some employers simply target an average number of labor hours per pound of product, and benchmark that between their own units or against industry averages. Unfortunately, such averages may have little relationship to the actual workload. One unit may be older or poorly maintained and require much more labor to achieve a given product output. Thus, the analysis of appropriate workload is very specific to a work situation, and the requirements will change over the life of a process.

16.4 REFERENCES

Hart, S. G. and Staveland, L. E. (1988). "Development of NASA-TLX (Task Load Index): Results of Empirical and Theoretical Research," *Human Mental Workload,* P. A. Hancock and N. Meshkati (Eds.) (North Holland: Elsevier Science Publishers), pp. 139-184.

Hill, S. G., Lysaght, R. J., Bittner, A. C., Jr., Bulger, J., Plarnondon, B. D. (1987) "Operator Workload (OWL) Assessment Program for the Army: Results from Requirements Document Review and User Interview Analysis" (Alexandria, VA: Army Research Institute for the Behavioral and Social Sciences.

Meister, D. (1985). *Behavioral Analysis and Measurement Methods.* (New York: Wiley).

Pew, R. W. and Mavor, A. S. (Eds.) (1998) "Modeling Human and Organizational Behavior," National Research Council Commission on Behavioral and Social Sciences and Education. (Washington, DC: National Academy Press).

Reid, G. B. and Nygren, T. E. (1988). "The Subjective Workload Assessment Technique: A Scaling Procedure for Measuring Mental Workload," *Human Mental Workload,* P. A. Hancock and N. Meshkati, eds. (North Holland: Elsevier Science Publishers), pp. 185–218.

Sanders, M. S. and McCormick, E. J. (1993), *Human Factors in Engineering and Design,* 7th ed. (New York: McGraw-Hill).

Sarno, K. and Wickens, C. (1995) "Role of Multiple Resources in Predicting Time-Sharing Efficiency: Evaluation of Three Workload Models in a Multiple-Task Environment," *The International Journal of Aviation Psychology, 5,* 107–130.

See, J. E. and Vidulich, M. A. (1998) "Computer Modeling of Operator Mental Workload and Situational Awareness in Simulated Air-to-Ground Combat: An Assessment of Predictive Validity," *International Journal of Aviation Psychology,* 8(4), 351–375.

Vidulich, M. A., Ward, G. F., and Schueren, J. (1991) "Using the Subjective Workload Dominance (SWORD) Technique for Projective Workload Assessment," *Human Factors,* 22 (6), 677–691.

CHAPTER 17

Shiftwork Issues

17.1 INTRODUCTION

Shift work can be defined as, "A work activity scheduled outside of normal daytime hours (an 8-hour period between 7 am and 7 pm) where there may be a hand over of duty from one individual or work group to another."

Badly managed shift work, especially night work and long working hours (overtime or double shifts) can lead to fatigue, which in turn leads to reduced vigilance, forgetfulness, impaired decision-making, impaired reaction times, poor communication and a general deterioration in mood and motivation. These effects are not permanent and can be reduced or eliminated when the person is fully rested. Fatigue can lead to operator errors and violations and is often a root cause of accidents. In the long term, effects can include: chronic fatigue, stomach disorders, cardiovascular disease, anxiety and depression. Fatigue affects a number of the key physical and mental abilities needed to carry out even fairly simple tasks safely. The implications for major accident hazard sites are obvious particularly with regards to operations (e. g. control panels misread) and maintenance staff (e. g. checks not completed prior to re-commissioning).

"After working approximately 29 hours straight, the last job I had to do was a simple engine component change, one I had done many times before. Following the fitment of the component, I could not focus on the correct rigging procedures. My concentration had lapsed to the point where I could not conduct a simple task." (Source: Reason & Hobbs, 2003)

Key facts about fatigue include:

1. Fatigue has biological causes.
2. The effects of sleep loss accumulate, leading to sleep debt.
3. If you ignore sleepiness, you will eventually fall asleep uncontrollably.
4. Two full consecutive nights of good sleep are needed for recovery.
5. Everyone needs about 8 hours sleep per night.

6. An internal body clock programs us to sleep at night and be alert during the day.

7. We are most ready to fall asleep around 3–5 am and 3–5 pm. The daily time for worst performance is 3–5 am. (See Figure 17-1)

8. The body clock does not, by and large, adapt well to night work.

9. The body clock is cued by changes between day and night, between work and home and by meals.

10. Day sleep is less restorative than night sleep and is typically lighter and shorter in duration.

17.2 TOOLS

A hierarchical approach should be adopted with preference given to the elimination of, or the reduction in the need for shift work. The following issues should be considered when evaluating and designing or re-designing shift systems:

- Shift times—workers perform best during the hours 7am–8pm. Move early starts forward from 6 am to 7 am.
- Shift duration—8-hour shifts, particularly at night, are considered to be ideal. A move to 12-hour shifts should be carefully assessed, particularly in safety-critical operations. Further overtime should only be permitted in exceptional circumstances. Control hours worked and overtime by keeping adequate records.

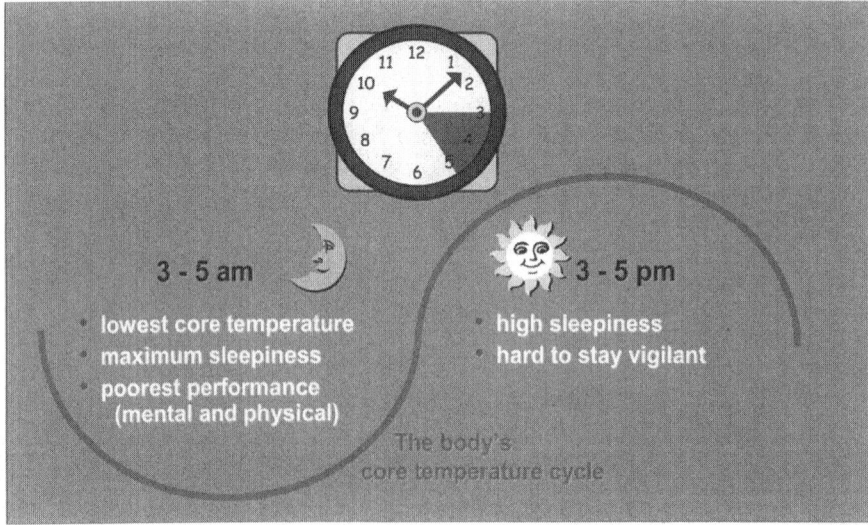

Figure 17-1: Time of day and effects on fatigue and performance

- Shift rotation—forward rotating shifts (morning, afternoons, then nights) are easiest for the body to adapt to. Avoid rotating shifts every 1—2 weeks.
- Breaks within shift—a quality break of 5-15 minutes every 1-2 hours is advocated.
- Breaks between shifts—allow a minimum of 12 hours to permit sufficient time for commuting to and from work (ideally less than an hour each way), meals and sleep. Plan some weekends off, advisably at least every 3 weeks.
- Napping—short, managed, naps at work can help to maintain alertness and performance.
- Rest days—ideally night shifts should be followed by at least two days off.
- Individual preferences—some people are more suited to shift work than others. People who prefer early starts are referred to as "larks" and those who prefer night shifts as "owls." Avoid permanent night shifts if at all feasible. Permanent adaptation is difficult because the normal light-dark cycle, meals and other social activities disrupt night workers internal clocks leading to chronic fatigue and poor performance.
- Staffing—provide adequate staffing levels and relief systems to avoid regular working of excessive hours or overtime. Put contingency plans in place to deal with an unexpected absence (do not overload everyone else). Consider establishing "sleep contracts" with employees who work shifts.
- Task management—restrict the number of safety-critical tasks planned for the night shift and rotate jobs at night to reduce levels of boredom and/or relieve physical fatigue. Arrange for a variety of interesting tasks to be performed at night and at other low points.
- Working environment—improve alertness by optimising lighting, temperature and comfort levels. Research indicates that bright light can aid adaptation to night work by resetting the body's internal clock. Avoid a warm stuffy environment that can cause drowsiness.

An initial risk assessment of the fatigue implications of different (rotating) shift patterns for safety-critical work can be made using a tool such as the Fatigue Index (FI) (HSE, CRR 254/1999). The Epworth Sleepiness Scale provides a measure of how likely an individual is to doze off and can be linked to a "sleep challenge" that individuals participate in, e.g. 80% of workers sleep score is below 8. The Owl–Lark (Morningness–Eveningness) Scale can be used by an individual to help them to identify where they are prone to fatigue through the day.

Other tools available include: Circadian Alertness Simulator (CAS), Fatigue Audit Interdyne (FAID), Sleep Activity Fatigue and Task Effectiveness models (SAFTE), Sleepwake Predictor, System for Aircrew Fatigue Evaluation (SAFE), and the Stanford Sleepiness Scale, information on which can be downloaded from the internet.

The organizational factors required to manage health and safety effectively can also be applied to shiftwork. Common factors in successful shiftworking businesses include:

- Accountability—someone has the responsibility for shiftwork management.
- Risk assessment and management—individuals at risk must be identified and risk factors evaluated. Careful planning of shift rosters, maximum hours of duty and recovery time is required.
- Participation—employees must be involved as early as possible in the design of or change to shift systems. Create a working group with representatives from relevant stakeholder groups.
- Education and training—raise awareness of managers and supervisors, employees, and their partners and or families of the problems associated with fatigue and also of the importance of sleep routines, breaks, nutrition, exercise and impact on family and social life. Check that the issues are understood and are being acted upon.
- Medical advice—must be accessible for those with existing medical conditions.
- Monitoring and review—examine accidents and incidents for evidence of fatigue and lessons learned. Go into the workplace and look for signs of employee fatigue, particularly in safety critical work. Set clear performance targets that enable you to assess the effectiveness of the changes you make, e.g. 10% reduction in Fatigue Index, 10% reduction in accidents and incidents, 10% increase in performance, etc.
- Culture—seek to develop a culture in which everyone accepts that fatigue is an issue and everyone works together to prevent it.

17.3 REFERENCES

Reason, J. and Hobbs, A. (2003), "Managing Maintenance Error: A Practical Guide" (Ashgate), ISBN 0 7546\1591 X.

HSE (1999), "Validation and Development of a Method for Assessing the Risks Arising from Mental Fatigue," CRR 254/1999 (Sudbury, UK: HSE Books), ISBN 0 7176 1728 9. Available at: http://www.hse.gov.uk/research/crr_pdf/2001/crr99254.pdf

17.4 ADDITIONAL REFERENCES

Step Change (2005), "Effects of Offshore Shift Patterns, OIM's Guidance for Offshore Rotas and Rest Periods" (Aberdeen: Step Change in Safety). Available on the Step Change website at: http://step.steel-sci.org

NIOSH (1997), "Plain Language about Shiftwork" (Washington, DC: National Institute for Occupational Safety and Health), Available online at: http://www.cdc.gov/niosh/

HSE (1990), "Reducing Error and Influencing Behaviour," HSG48 (Sudbury, UK: HSE Books), ISBN 0 7176 2452 8.

OSH (1998), "Stress and Fatigue: Their Impact on Health and Safety in the Workplace"

(Wellington, NZ: Occupational Safety and Health Service of the Department of Labour). Available online at: http://www.osh.dol.govt.nz/order/catalogue/pdf/stress.pdf.

Barton, J., Spelten, E., Totterdell, P., Smith, L., Folkard S. and Costa, G. (1995), "The Standard Shiftwork Index: a Battery of Questionnaires for Assessing Shiftwork-Related Problems," Work & Stress vol. 9, no. 1, pp. 4–30.

CHAPTER 18

Manual Materials Handling

18.1 INTRODUCTION

Manual materials handling refers to the lifting, lowering, pushing, pulling, carrying or moving of a load (e.g. tools, covers, equipment, process consumables, etc.) by hand or by bodily force.

Operators experiencing discomfort (stresses and strains) associated with manual materials handling are more prone to making errors, which may lead to an undesired process event.

> A 55-gallon drum of catalyst was meant to be kept in refrigerated storage. Rather than walk the long distance each time catalyst was needed, operators kept an opened drum adjacent to the reactor. The catalyst degraded in the warm environment, and several bad batches had to be dumped to an outdoor pit, causing a serious environmental incident and a potential fire.

Poor manual materials handling can lead also to a range of conditions, known as work-related musculoskeletal disorders (WRMSDs) or musculoskeletal disorders (MSDs), that affect the muscles, tendons, nerves and supporting structures of the musculoskeletal system. These disorders are regarded as cumulative in origin arising from prolonged and repeated exposure to excessive levels of both physical and psychosocial stressors (are those things that may influence a person's response to their work and workplace conditions, including their relationship with their supervisor and colleagues) at work. Individual size, strength and joint alignment; and non-occupational activities such as sports and hobbies can also increase risk of injury (HSE, 2004).

"More than a third of all over-three day injuries reported each year to HSE and local authorities are caused by manual handling."

Good manual materials handling can improve performance while cutting costs, incidents and injuries.

An effective manual materials handling program requires senior management commitment, clear roles and accountabilities for those responsible for their prevention, full employee participation, effective communication and management of change processes, and the use of competent people when examining problems and identifying what can be done to prevent or control them. An effective framework for managing the risks associated with manual materials handling is shown in Figure 18-1.

18.2   MANUAL MATERIALS HANDLING GUIDELINES

Address all lifting, lowering, pushing, pulling, carrying or moving tasks relating to operations and maintenance under all modes of operation including emergency situations. Consult with your employees—they undertake the work and will be able to

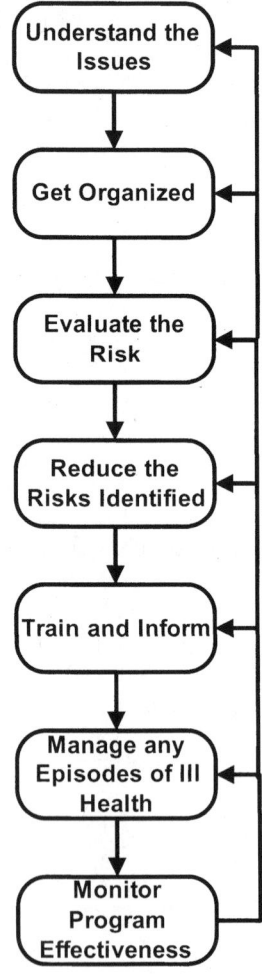

Figure 18-1: A framework for managing MSDs in the workplace (modified from HSE, 2002)

help you to identify problem areas. Focus on the key risk factors: task, load, working environment, individual capability and work organisation. A simple checklist that can be used to help identify whether or not you have a problem is given below.

1. Are there any medically diagnosed cases of MSDs at your workplace?
2. Do employees complain of aches or pains when performing certain tasks?
3. Are there any jobs which employees are reluctant to perform or regularly request to be taken off?
4. Are there any tasks that involve large vertical movements or very long carrying distances?
5. Do employees work in postures that are awkward and/or held in fixed positions for prolonged periods of time? The more joints deviate from their neutral position the greater the risk.
6. Are there any tasks that involve repetitive handling for prolonged periods?
7. Are there any tasks that require strenuous pushing or pulling?
8. Are employees required to work with loads that are unwieldy, unstable or difficult to grasp?
9. Do employees perform heavy manual handling? The NIOSH lifting equation specifies a lifting limit of 23 kg but this is reduced for unfavorable lifting conditions.
10. Do employees work in a poorly lit and or particularly cold/hot/humid work environment?
11. Do employees have to work to tight deadlines?
12. Do employees have poor control over work and working methods?

If you've answered yes to one or more of the questions then a more detailed evaluation may be required. A range of tools are available that you can use to evaluate and design or re-design problem tasks. Some of these tools are listed in Table 18-1. Manual material handling evaluations should be re-evaluated whenever a significant change or reportable injury occurs.

Strong leadership from management and active employee participation in the identification and subsequent implementation of remedial measures is vital. A hierarchical approach should be adopted with preference being given to elimination of risk at the source through automation or mechanisation, although these items still have to be maintained. If elimination is not possible introduce mechanical aids (assisted lifting devices), then consider individual risk factors and their inter-relationship with a view to designing tasks that fit the individual. Consider the following general approaches:

- Redesign the load, e.g. lighter, smaller, easier to grasp, etc.
- Redesign the task, e.g. reduce twisting and over-reaching, reduce work at floor level or above shoulder height, ensure adequate rest periods, vary work, keep loads close to body, etc.

Table 18-1: Manual materials handling analysis tools

Tool	Use	Source
Manual Handling Assessment Charts (MAC)	Quick risk assessment tool for lifting and lowering, and carrying.	Health and Safety Executive HSE (2003)
NIOSH Lifting Equation	Quantitative analysis of lifting with several constraints	Waters, T., Putz-Anderson, V., and Garg, A. (1994).
Push/Pull/Carry Calculator	Estimates the suggested maximum force that can be applied during pushing and pulling and the weight that can be carried.	WorkSafe British Colombia www.healthandsafetycentre.org/ppcc/default.htm
3D Static Strength Prediction Program (3D SSPP)	Predicts static strength requirements for tasks such as lifts, presses, pushes, and pulls. Data comparisons with NIOSH guidelines can be made.	University of Michigan, Centre for Ergonomics www.engin.umich.edu/dept/ioe/3DSSP/
Weight Limit Tables	Defines weight limits for female and male lifting, lowering, pushing, pulling, carrying tasks.	Snook, S. H. and Ciriello, V. M. (1991).

- Redesign the work environment, e.g. improve access, remove space constraints, improve lighting, improve thermal conditions, reduce exposure time to vibration, etc.
- Improve work organization, e.g. engage employees, enhance communication, match tasks to employee skills, promote teamwork, etc.

Formal manual material handling studies should be conducted during facility design and again during major modifications. These studies provide designers and operators with the greatest opportunity to eliminate the need for manual handling once a facility is in operation. Such studies can also have a significant impact on plant layout, subsequent ease of maintenance, and on the selection of lifting equipment.

Employees should be provided with the necessary skills required to conduct the job, and retrained periodically to re-enforce safe working practice. Guidance should also be given on the steps that can be taken to reduce the occurrence of awkward postures, minimize the levels of force applied, and reduce repetition. Training must be focused, relevant, and capable of being put into immediate practice. A wide range of resources can be accessed through the websites operated by national health and safety regulatory bodies such as the United States Occupational Health and Safety Administration and the European Agency for Safety and Health at Work.

Establish management systems in place which:

- Encourage early reporting of any health problems. This will usually avoid any serious problems developing.
- Provide appropriate advice for users who report health problems. This might range from reassurance to advice on risk factors through to completion of a postural checklist.
- Provide for referral to a health professional for early diagnosis and treatment, implementation of tailored work programs, managed treatment regimes, and planned return to work.

Systems should be established to monitor and review the effectiveness of your manual materials handling program. Monitoring is the ongoing and regular appraisal of the systems and procedures to control manual materials handling risk. There are two broad approaches—active and passive monitoring. Passive approaches, for example, would monitor staff complaints, sickness absence and symptoms reported. Active systems comprise task analysis, health examinations/interviews and walk-throughs. Review is less frequent, more strategic, and should assess the overall effectiveness of the manual materials handling program. It should be undertaken when monitoring indicates that the current program is not providing adequate control of workplace manual materials handling risk. Ideas for improvement should be captured and tracked through to completion.

If the exothermic reaction of a polymer started to go out of control (an infrequent event) manual addition of a shortstop by the operator was made through a vessel on top of the reactor on the 3rd floor of the building. The shortstop was stored in a refrigerator close by. Normally an operator was always present on the third floor and would either know through local alarms and instrumentation when shortstop was needed or received a call from the control room which was somewhat remote. The plant underwent staffing reductions which eliminated the third floor operator. When the reactor did subsequently go out of control after the change was made there wasn't enough time to send one of the remaining operators to the third floor to add the shortstop and the runaway reaction ensued resulting in a release to a catchpot.

18.3 REFERENCES

HSE (2002), "Upper Limb Disorders in the Workplace," HSG60 (Sudbury, UK, HSE Books), ISBN 0-7176-1978-8.

HSE (2003), "Manual Handling Assessment Charts (MAC)," Leaflet INDG383 (Sudbury, UK: HSE Books).

HSE (2004), "Getting to Grips with Manual Handling: A Short Guide," INDG143 (rev2) (Sudbury, UK: HSE Books), ISBN 0 7176 2828 0.

Snook, S. H. and Ciriello, V. M. (1991). "The Design of Manual Handling Tasks: Revised Tables of Acceptable Weights and Forces," *Ergonomics, 34*, p. 1197.

University of Michigan, Centre for Ergonomics, www.engin.umich.edu/dept/ioe/3DSSP/

Waters, T., Putz-Anderson, V., and Garg, A. (1994). "Application Manual for the Revised NIOSH Lifting Equation, Publication No. 94-110 (Washington, DC: US Department of Health and Human Services).

WorkSafe British Colombia, www.healthandsafetycentre.org/ppcc/default.htm

18.4 ADDITIONAL REFERENCES

NZ OSHS (2001), "Code of Practice for Manual Handling" (Wellington, New Zealand: Occupational Safety and Health Service of the Department of Labour and the Accident Compensation Corporation). Available online at: http://www.osh.dol.govt.nz/order/catalogue/pdf/manualcode.pdf

Worksafe (2005), "Code of Practise for Manual Handling" (West Perth, Australia: Western Australia Department of Consumer and Employment Protection). Available online at: http://www.safetyline.wa.gov.au/pagebin/codewswa0221.pdf.

American Bureau of Shipping (2003), "Guidance Notes for the Application of Ergonomics to Marine Systems" (Houston, TX: American Bureau of Shipping). Available online at: http://www.eagle.org/rules/downloads/86-Ergo.pdf

USEFUL WEBSITES

The websites listed below all provide access to a range of information, including guidance on diagnostic techniques, and tools that you can use to help evaluate and manage workplace manual materials handling risk.

Australia, WorkSafe Western Australia Webpage: http://www.safetyline.wa.gov.au

Canada, WorkSafeBC Webpage: http://www.worksafebc.com

European Union, European Agency for Safety and Health at Work Webpage: http://osha.eu.int/

United Kingdom, Health and Safety Executive Webpage: http://www.hse.gov.uk/

United States, Occupational Health and Safety Administration Webpage: http://www.osha.gov/

United States, The National Institute for Occupational Safety and Health Webpage: http://www.cdc.gov/niosh/homepage.html

Management Systems

CHAPTER 19

Safety Culture

19.1 INTRODUCTION

Most organizations with safety-critical operations recognize the value of developing a strong safety culture. An organization's culture radically influences the way people at all levels think and behave at work, and thus affects the occurrence or absence of accidents, process safety incidents and poor performance. However, it is difficult to assess your own safety culture without a suitable framework and process to measure it, and compare with the ideal. This chapter defines safety culture, outlines why it is important and describes some safety culture models, along with methods of analyzing and then improving the organization's safety culture.

Safety culture is now recognized as a significant element of human factors. It is seen as critical to improving performance due to the findings from investigations into major disasters in the process industries (e.g. Flixborough and Piper Alpha), other industries such as nuclear power (e.g. Three Mile Island and Chernobyl) and transportation (Exxon Valdez and Space Shuttle). All of the investigations into these accidents concluded that systems broke down catastrophically despite the provision for complex technical safeguards. These disasters were not primarily caused by engineering failures, but by the action or inaction of the people designing, managing, maintaining and/or operating the system. Taking inappropriate risks, not following procedures and a belief that "productivity is the most important thing in our business" are all indicators of a weak safety culture which invariably negates the benefits that good engineering, procedures, training and management systems provide.

A plant manager made a determined effort to achieve new production levels. Unfortunately this sent a message to the supervisors and workers that production was more important than anything else—a conflicting priority. Several safety compromises were made and an accident resulted which injured several workers and shut down the plant for many weeks.

19.2 WHAT IS SAFETY CULTURE?

The term *safety culture* was defined by the International Atomic Energy Agency in their report on the Chernobyl nuclear power plant disaster which occurred in 1986

(IAEA, 2005). The errors and violations of operating procedures which contributed to the Chernobyl disaster were seen by some as being evidence of a poor safety culture at the plant (Lees, 1996). This led to a number of studies which attempted to measure safety culture in a variety of different high-risk, high-hazard industries.

The UK Advisory Committee on the Safety of Nuclear Installations produced the most widely accepted and comprehensive safety culture definition (HSE, 1999):

"the product of individual and group values, attitudes, perceptions, competencies, and patterns of behavior that determine commitment to, and the style and proficiency of, an organization's health and safety management. Organizations with a positive safety culture are characterized by communications founded on mutual trust, by shared perceptions of the importance of safety and by the efficacy of preventive measures."

An earlier more tangible definition which many can relate to was offered by Uttal (1983):

Culture (of an organization, of which safety culture is part) is shared values (what is important) and beliefs (how things work) which interact with an organization's structure and control systems to produce behavioral norms (the way we do things around here).

The characteristics of a positive safety culture are presented in Table 19-1. These were determined by combining the characteristics of low accident organizations using safety culture/climate research. It is worth noting and should be of no surprise

Table 19-1: Features associated with a positive safety culture

1. Hardware:
   - Good plant design, working conditions and housekeeping
   - Perception of low risk due to confidence in engineered systems
2. Management systems:
   - Confidence in safety rules, procedures and measures
   - Safety prioritized over profits and production
   - Satisfaction with training
   - Good job communication
   - Good organizational learning
3. People:
   - High levels of employee participation in safety
   - Trust in workforce to manage risk
   - High levels of management safety concern, involvement and commitment
4. Behavior:
   - Acceptance of personal responsibility for safety
   - Frequent informal safety communication
   - Willingness to speak up about safety
   - A cautious approach to risk
5. Organizational climate factors:
   - Low levels of job stress
   - High levels of job satisfaction

that organizations strong in these areas are successful in all aspects of business performance (PRISM, 2005).

19.3 TOOLS

The assessment of safety culture requires a suitable framework in which to measure and compare the findings with an ideal model. Numerous safety culture and safety attitude surveys are available and have been used with varying degrees of success. Some industry HSE experts likened several of these to "describing the water to a drowning man"—in other words, they may in an intellectual way describe the nature of the problem, but offer little practical help. To avoid these pitfalls, as industry has become more informed, it expects certain characteristics from these tools and from the specialist consultants which offer them, namely:

- A learning experience for all participants through a practical and intuitive approach.
- Involves all levels of staff, especially employees and contractors at the workfront.
- All participants contribute in an uninhibited way to speak freely about the positives and negatives (by sharing examples) of "what it's like to work here" in the context of the safety culture model chosen.
- Action focused, supporting the development of the short and medium term HSE development plans.
- Resulting recommendations will be specific to the team represented at the workshops.
- Acknowledges that cultural development is a journey and that organizations are where they are without judging them good or bad.
- Includes a feedback process for all participants

There are a number of models developed in recent years that describe how an organizations safety culture evolves over time and with the appropriate nurturing. The International Association of Oil & Gas Producers (OGP) published a paper (www.ogp.org.uk) which includes a HSE Culture Ladder model (Figure 19-1) as an example of this. The figure was derived from the following project: *PRISM: Process Industries Safety Management Thematic Network,* which can also be accessed through the website.

Another contribution is "Hearts and Minds" an example of a comprehensive toolkit which has been developed by Royal Dutch/Shell Group and implemented within its companies (www.energyinst.org.uk/heartsandminds). Its aim is to develop organizational culture with a stated belief that: "*A solidly implemented HSE management system is an essential basis for good HSE performance. Outstanding performance and continuous improvement will only be achieved when there is a culture in which the elements of the management system can flourish*". The model underpinning the toolkit is the HSE culture ladder model referred to earlier in this section and illustrated in Figure 19-1.

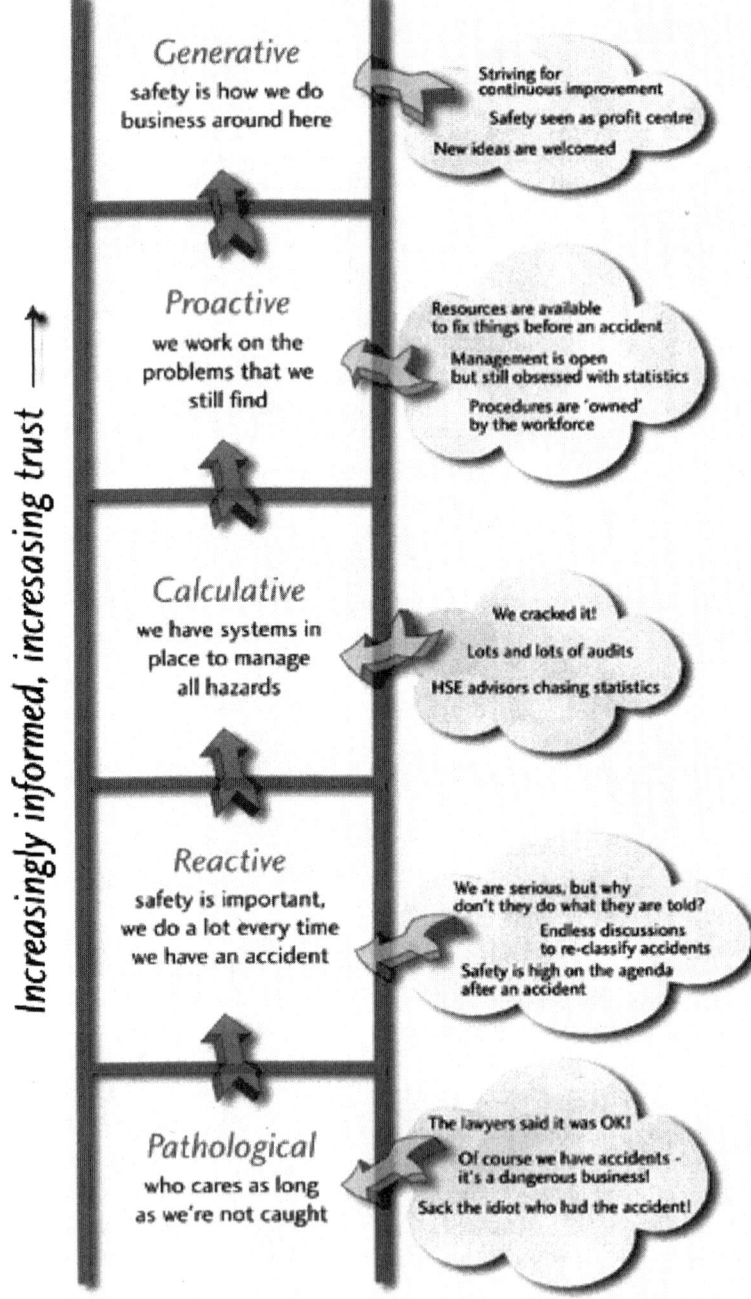

Figure 19-1: OGP's HSE Culture Ladder (PRISM, 2005).

The UK oil sector and the UK's Health and Safety Executive recognized that the failure of local safety initiatives, which had worked successfully and improved performance on another plant (even within the same company), was often due to the lower level of cultural maturity of that operation. They worked with The Keil Centre to develop the Safety Culture Maturity Model® (HSE, 2001) which is described in Figure 19-2. This model was then further enhanced into a practical method of assessment and the identification of improvement opportunities. This model involves short, facilitated workshops at an early stage, where representatives of the main occupational groups openly select the maturity level they believe the organization is at. They then suggest practical actions they and others could take to move forward in the 10 elements described in Table 19-2. These elements were derived from existing safety culture and human factors literature and from industry experience. By design, the model meets all the characteristics referred to earlier in this section and has been used successfully throughout the world, across a number of companies and industries.

Many organizations now use these tools to help them determine improvement techniques which will complement their current cultural maturity level—a particular behavioral safety program may not be as effective without strong visible

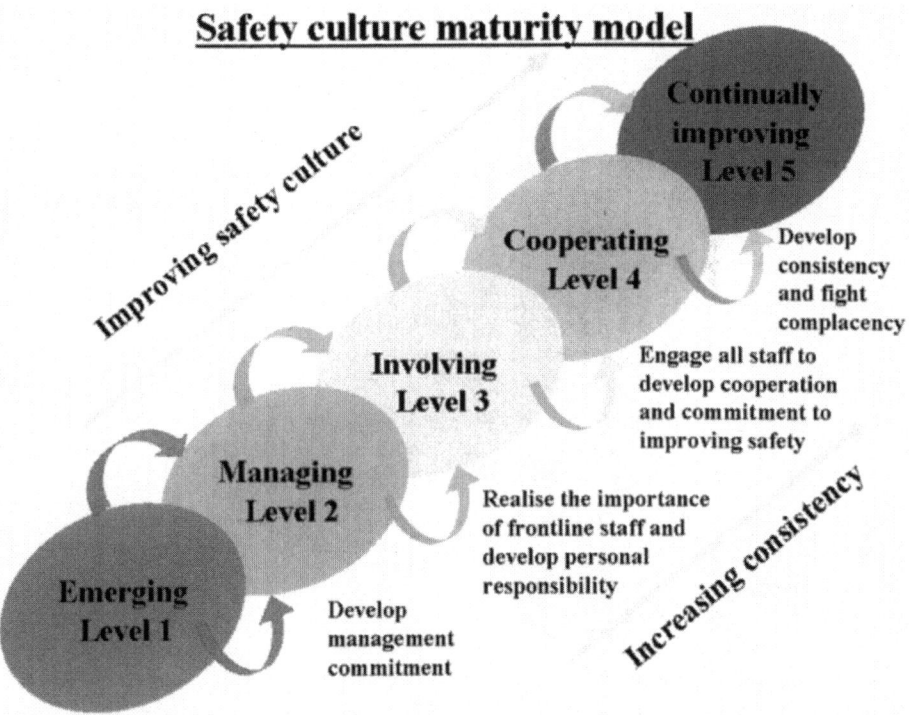

Figure 19-2: Safety Culture Maturity Model®

Table 19-2: The SCMM® includes ten key elements of safety culture derived from existing safety culture and human factors literature, and industry experience.

• Visible management commitment	• Risk taking behavior
• Safety communication	• Trust between management and front-line staff
• Learning organization	• Industrial relations and job satisfaction
• Health and safety resources	• Competency
• Participation in safety	• Productivity versus safety

management commitment, trust or evidence of a learning organization. Furthermore, as a safety culture matures, further improvement does not necessarily involve "more of the same." In short, the complexity of any safety culture improvement initiative must be matched to the "maturity" of the existing safety culture to ensure success. The type of safety culture improvement method needed to support safety culture development changes as it matures. This development concept can be compared with child development—the types of actions necessary to help an infant learn, develop and mature differ from those appropriate for a young adult.

19.4 SAFETY CULTURE: A PROCESS INDUSTRY CASE STUDY

A successful safety culture improvement project was conducted in 2000 at a major oil company site in Scotland (Joyner, 2001). The site was regarded as having a relatively strong safety culture and acceptable levels of safety performance. However, management recognized that unsafe behavior was a problem as it was a basic cause of their incidents. They wanted to get to the root cause and decided to first engage the workforce and then identify realistic, practical actions to further enhance their safety culture and business performance. The Safety Culture Maturity Model* helped senior managers plan and design a safety culture improvement initiative appropriate to their local needs and circumstances which had the buy-in of the workforce.

The assessment found that the maturity of most of the ten safety culture elements were at level 3 "involving," and moving towards level 4 "cooperating." The safety culture was therefore perceived by the team to be relatively mature. However, differences in levels were found between major occupational groups, with the maintenance and marine teams assessing their maturity level to be lower. Since the safety culture a worker experiences shapes the way he/she thinks and behaves, it makes this an important finding. Improvements were then tailored to the needs and maturity of the group, rather than adopting a "one size fits all" approach. All occupational groups were able to identify practical actions to improve the maturity of their safety culture.

*Safety Culture Maturity is a registered trademark of The Keil Centre Ltd.

19.5 BENEFITS

The local manager reviewed the project several months after completion. At the organizational level the following actions had taken place:

- Site staff had designed and implemented their own simple near-miss reporting system, which quickly resulted in the reporting and correction of unsafe conditions. These conditions had existed for a number of years.
- More face-to-face communication (formal and informal) was occurring between management and work teams.
- Regular shift team leader meetings had been re-introduced to discuss process and safety improvements. The teams "livened up" the meetings by inviting guests to talk on production and/or safety issues.
- Shift and maintenance safety teams, which had been dormant for a number of years, had been revitalized by the team members themselves. These teams now realized that they could make a difference to safety and were taking ownership for important safety improvement actions.
- Safety issues which had been reported were now being consistently "closed out."
- Increased recognition was being provided to members of staff who made a positive contribution to improving safety.

It also became apparent that an additional educational benefit had been realized—workshop participants learned about the nature of safety culture, its main components, and how they could personally contribute to enhancing its maturity. The local manager observed "people now possess a clearer understanding of the importance of their individual behaviour in improving safety performance, and demonstrate a greater willingness to take ownership of improvement actions".

19.6 REFERENCES

HSE (1999), "Development of a Business Excellence Model of Safety Culture" (London, UK: U.K. Health and Safety Executive).

HSE (2001), "Safety Culture Maturity Model" (London: U.K. Health and Safety Executive) downloadable at http://www.hse.gov.uk/research/otopdf/2000/oto00049.pdf.

IEAA (2005), Web site materials on Chernobyl at www.ieaa.org.

Joyner, P. (2001), "Towards a Mature Safety Culture," *Proceedings of IChemE Hazards Symposium* No. 148, Paper 49 (Rugby, UK: The Institution of Chemical Engineers).

Lees, F. P. (1996), "Loss Prevention in the Process Industries," Appendix 22: Chernobyl (Amsterdam, NE: Elsevier).

PRISM (2005), "Safety Culture Application Guide for the European Process Industries" (Warwickshire, UK: European Process Safety Center), www.prism-network.org.

Uttal, B. (1983), The corporate culture vultures, *Fortune Magazine,* 17th October.

19.7 ADDITIONAL REFERENCES

Cooper, D. (1997), "Improving Safety Culture: A Practical Guide" (Hoboken, NJ: Wiley).

Florczak, C. (2002), "Maximizing Profitability with Safety Culture Development" (Amsterdam: Elsevier).

NSC (1999), "Safety Culture and Effective Safety Management" (Itasca, IL: National Safety Council).

Roughton, J. (2002), "Developing an Effective Safety Culture: A Leadership Approach" (Amsterdam: Elsevier).

CHAPTER 20

Behavior Based Safety

20.1 INTRODUCTION

Prior to the 1990's, most safety professionals and government regulators placed all of their focus on the elimination of hazardous conditions to prevent workplace injuries (DuPont, 1990). Well-known business guru and educator Peter Drucker (Drucker, 1993) states that "The Occupational Safety and Health Administration (OSHA) operates on the assumption that an unsafe environment is the primary cause of accidents." Even though this has resulted in great strides in the elimination of hazardous conditions, workplace injuries continue to occur at a steady and predictable pace. This has prompted some researchers to argue that engineering and management fixes alone are not enough.

Recent research data indicate that at-risk behavior is the root cause of 85 to 90% of all workplace injuries (Krause, 1999). Because of this revelation, many large and respected organizations are now looking at adding behavioral based safety (BBS) to their arsenal of tools to prevent workplace injuries. For example, one of the largest tire manufacturers recently announced that they are shifting 80% of their safety efforts to preventing at-risk behavior (Porter, 1999).

BBS can best be described as workers looking after each other. BBS is a method of sampling behavior (through observation) in the workplace, and comparing what was observed against a specification of safe performance. After an observation, feedback is used to reinforce all observed safe behaviors, encouraging them to continue. Constructive feedback then takes the form of structured two-way discussions in which the observer and the performer identify and understand the root cause of any observed at-risk behavior and to provide corrective actions to promote safe behavior. As with other quality control processes, BBS requires behavior sampling, data review, and control strategies to be effective.

Behavior is neither good nor bad (even though as a child, the word usually preceded punishment). The true definition of a behavior is simply "an observable act." Quality gurus will tell us that if you can see it you can measure it and if you can measure it you can manage it.

In the world of Behavioral Based Safety, behaviors are seen as either "safe" or "at-risk." The big question is, why do workers take risks? This question is asked repeatedly, especially after an incident. For example, why would an experienced chemist pour acid without wearing a face shield and chemical resistant gloves?

What would motivate a 20-year veteran employee to enter a clearly labeled confined space without taking the necessary precautions?

Current literature strongly indicates that consequences (what happens after a behavior) is the driving force. This key premise can be traced back to the works of noted behavioral scientist, B. F. Skinner, who explained that " the reinforcing effects of things are the province of behavioral science" (Topf, 1995). Research indicates that the reason people continue unsafe behavior, regardless of knowledge, is because of the "positive, immediate and certain consequences associated with the unsafe behavior" (ShamRao, 1999).

Before Skinner's research, behavior was typically thought to be motivated by the antecedent or trigger. However, as B. F. Skinner points out, consequences are "more powerful determinates of behavior than are antecedents" (Krause, 1990). Krause (1990) continues by stating that:

> This relationship can be best illustrated by using the simplistic example of a ringing telephone. The phone rings (the antecedent) and we pick it up (the behavior) and we speak to the person calling (the consequence). Conventional wisdom led us to believe that the ringing of the phone (the antecedent) was the motivating factor behaving the behavior of picking up the phone. But, behavioral science taught that the consequence, of speaking to the person who called, is in actuality the strongest motivator of the behavior of picking up a ringing phone.

The power of the consequence was demonstrated by asking how many times one would continue to pick up a ringing phone if there were no one on the other end—the average person would not continue to pick up the phone. Consequences can either increase behavior or decrease behavior. The manner in which consequences compete for control of behavior is based on three factors; significance, timing and probability.

A plant site was not particularly concerned about good housekeeping. As a result, a stain on the side of a reactor due to sloppy vessel charging remained in place. Unfortunately, a real leak eventually developed in the reactor and it went unnoticed for a while since it was not visible on top of the existing stain.

20.2   TOOLS

ABC Analysis considers Antecedents, Behavior and Consequences. Antecedents include triggers of behaviors that prompt action as well as constraints that shape action, indicating what is and what is not behavior. Behaviors consider the behaviors that are required to complete the objectives. Behaviors lead to consequences, which may be positive, negative or sometimes a combination of both (changingminds.org).

Consider the example of a worker not wearing hearing protection in a high noise area.

ABC Analysis is accomplished by first listing the behavior to be analyzed. For the hearing protection example the behavior is "not wearing hearing protection." Next, in a non-threatening manner, the affected population, or a sample of that population, is surveyed to determine what antecedents trigger the behavior. In the example, it was determined that the antecedents were: "that the protection was not readily available," "the hazard was not understood," "the employees were in a hurry," and that the "supervisor of the area did not wear hearing protection." These are listed in Table 20-1.

The next step is to determine and list the consequences of the behavior. In the example, the consequences of not wearing hearing protection are: greater comfort, the employee does not have to search for hearing protection, a saving of time, and unfortunately, hearing damage.

Once the antecedents and consequences are determined and listed the consequences are analyzed to determine their significance, timing and probability of the consequences. Significance is denoted as either of "P" for positive or "N" for negative. In the case of not wearing hearing protection the consequences of greater comfort, not having to search for the protection and the saving of time are all determined to be positive consequences.

Next, the consequences are analyzed for timing. Consequences that are considered to occur immediately following a behavior are denoted with an "I." Consequences considered to follow later are denoted with an "F," for future. For our example, the three consequences that were positive also were determined to occur immediately following the behavior.

Finally, the probability of the consequences were considered. Those that are perceived to be certain to follow the behavior were labeled with a "C." Those that are perceived as not likely, or uncertain to occur are marked "U."

The results of the ABC Analysis for the hearing protection example are summarized in Table 20-1.

If hearing damage is a consequence of not wearing hearing protection, then why doesn't everyone wear protection every time they are in "loud" areas? Consequences

Table 20-1: ABC Analysis example

Antecedent	Behavior	Consequence	P/N	I/F	C/U
Protection not easily available	Not wearing hearing protection	Greater comfort	P	I	C
Did not understand hazard		No searching for hearing protection	P	I	C
In a hurry		Saves time	P	I	C
Supervisor does not wear hearing protection		Damage to hearing	N	F	U

Adapted from Daniels, 1994.

that are positive "P," immediate "I," and certain "C," are more likely to increase a be-havior (ShamRao, 1999). To the worker, greater comfort, not having to search for the protection and saving time are more powerful drivers of behavior than the negative "N," future "F" and uncertain "U" consequence of damage to hearing.

Before beginning a behavioral based safety process, many practitioners recom-mend first understanding how the employees perceive the current safety efforts. This understanding allows the organization to assure that any perceived shortcom-ings in the traditional focus on conditions are addressed before attention is turned to behavior. The most common tool to accomplish this is the comprehensive safety culture assessment (HSE, 1999). Effective assessments will also identify where to place efforts and resources to overcome organizational obstacles to a safe work-place. See Chapter 19 on Safety Culture.

The following steps show how a BBS tool is applied.

Step 1

A behavior based safety process is usually done under the direction of a consultant or qualified facilitator. First, select a "steering team" comprised of members best determined to assure the success of the process. This team is often comprised en-tirely of hourly employees and, with management support and resources, is driven from the bottom up rather than from the top down. This team then starts the process by analyzing the last few years of injury data to determine the top categories and types of injuries. Next, the data is further analyzed to determine what "at-risk" be-haviors led to those injuries and to prioritize these behaviors. The team then writes definitions that are the direct opposite of these behaviors. In other words, the defin-itions describe the "safe" behavior that, if performed would prevent injuries. De-scriptions of safe behaviors are written so that they are observable. These behaviors are then listed in a checklist which the workers can use to remind them of the be-haviors to look for and monitor.

Step 2

Next, the steering team learns how to make observations using the definitions and checklist in a non-threatening manner, always without names or any attached disci-pline. They also focus on how to provide effective feedback and to use problem solving tools to develop and track corrective actions. Once the team is comfortable and confident, they begin the process by introducing BBS to the entire site, or at least to all of the affected employees. The presentation usually includes an overview of BBS and some demonstrations of observation, feedback and problem solving. The presentation is also used to reassure the workforce of anonymity and to answer questions and concerns.

Step 3

Finally, the team selects and trains about 20% of the affected workforce as ob-servers. Information collected by the observers is captured on the standardized checklists so that it can be tracked and analyzed. Descriptions of safe behaviors are written so that they are observable. These behaviors can be listed in a checklist which the workers use to remind them of what behaviors to look for and monitor.

Many organizations enter this data into their own or commercially available databases for easy manipulation and analysis. Compiled data is used by the steering team to further problem solve and to take corrective actions and track progress. The most sophisticated organizations track the number of observations per area or task and eventually are able to show the relationship between the observations and decreased injuries.

20.3 EXPECTED RESULTS

Behavior is necessary to accomplish any task. Therefore, the principles discussed above can be applied to reinforce and improve performance for any desired outcome, including: safety, quality, athletic performance, artistic performance and of course the topic of this publication, process safety. Behaviors critical to not only assuring process safety, but also in providing ongoing compliance documentation are easily defined, observed and reinforced.

Success of any behavioral based safety process is contingent upon how well the employees who conduct the observations and provide the feedback are able to coach the workers they observe. Successful "coaches" learn how to provide feedback, which promotes the continuation of safe work while engaging in two-way conversations, assuring the development and tracking of corrective actions. What happens next is truly amazing. These one-on-one, structured safety discussions change the safety culture so that the workers look after each other, and work to prevent injuries.

20.4 REFERENCES

Daniels, A. (1994) "Bringing Out The Best in People" (New York: McGraw-Hill).

Drucker, P. (1993) "Managing in a Time of Change" (New York: Truman Tally Books).

DuPont (2005), "Safety Training and Observation Program (STOP)" (Wilmington, DE: E. I. DuPont de Nemours and Co.). Information available at: http://www.dupont.com/stop/en/

HSE (1999), "Development of a Business Excellence Model of Safety Culture" (London, U.K.: U.K. Health and Safety Executive).

Krause, T. (Ed.) (1999) "Current Issues in Behavior-Based Safety" (Ojai, CA: Behavioral Science Technology, Inc.).

Porter, M., "Successes and Challenges of a Large Organization Symposium," *Safety Leadership Conference,* Del Ridder (Chair), Conducted at the meeting of the Summit County Safety Council, Akron, OH.

ShamRao, A. (1999) "Whole Safety Systems" (Chicago: Divaker Ltd.)

Topf, M., and Petrino (1995) "Change in Attitude Fosters Responsibility For Safety." *Professional Safety,* 43, 24–27.

20.5 ADDITIONAL REFERENCES

Geller, E. S. (2001), "Keys to Behavior-Based Safety" (Washington, DC: Government Institutes).

Krause, T. R. (1996), "The Behavior-Based Safety Process: Managing Involvement for an Injury-Free Workplace," 2nd ed. (Hoboken, NJ: Wiley).

Krause, T. R. (Ed.) (1999), "Current Issues in Behavior-Based Safety" (Ojai, CA: Behavioral Science Technology, Inc.).

McSween, T. E. (2003), "The Values-Based Safety Process: Improving Your Safety Culture with Bahavior-Based Safety" (Hoboken, NJ: Wiley-Interscience).

CHAPTER 21

Project Planning, Design, and Execution

21.1 INTRODUCTION

The inclusion of human factors in capital projects has been shown to be good business (Robertson, 1999). Human factors, if considered early in the project, can be extremely cost-effective—a project costs no more to design and build while considering the human during the design.

Human factors is typically built into projects to improve the design of equipment and facilities. The benefits of including human factors in project planning, design and execution include:

1. Reducing the potential for injuries
2. Preventing the potential for damage to existing equipment and facilities,
3. Improving the cost-effectiveness of the design (Attwood and Fennell, 2001) by reducing the costs due to design changes and retrofitting.
4. Reducing the potential impact of "human error" as the project is commissioned and operated

Projects are conducted every day in all process plants. Most projects are typically not major capital projects, but smaller projects, often called "local" or "base" projects or just plant repair and modifications. The manner in which major and local projects are conducted profoundly influences how human factors is considered in each type of project and the design of the tools that are used. It is important to remember that not all "projects" are created equal.

21.2 HUMAN FACTORS TOOLS FOR PROJECT MANAGEMENT

Figure 21-1 illustrates the human factors tools that can be used during each phase of the planning, development and design process. Each is explained in Table 21-1. Some of the tools are stand-alone, such as LINK analysis. Others are integrated into existing key project tools such as Process Hazard Analyses (PHAs). Chapter 9 in Attwood et al. (2003) provides a more detailed description of each tool in Table 21-1.

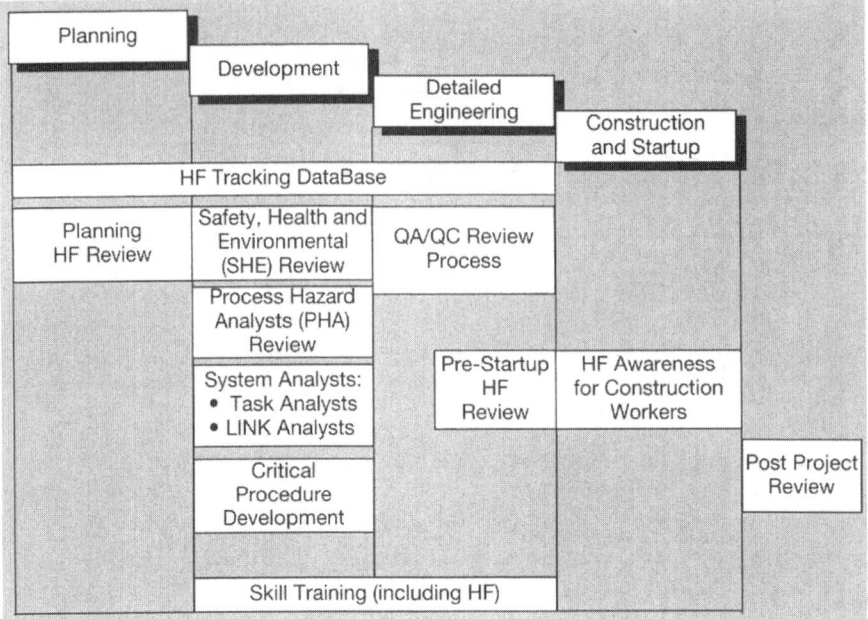

Figure 21-1: Human factors tools in the life cycle of a chemical plant. [Adapted from Attwood et al. (2003).]

Table 21-1: Description of human factors tools from Figure 21-1

HF Tool	Purpose
Human Factors Tracking DataBase	The Human Factors Tracking DataBase (HFTDB) consists of a standard spreadsheet that captures the Human Factors Design Issues as they are identified. It is initiated by the project coordinator and handed off to successive project team leaders as it passes from one phase to the next.
Planning HF Review	The list of "screening questions" is intended to identify basic Human Factors information about the proposed plant. As issues are identified, they are tracked on the HF Tracking DataBase. Example questions are listed in Table 21-2.
SHE Review	This review is conducted to identify and address key safety, health and environmental (SHE) hazards that could harm people, damage property and adversely affect the process. The objective of the SHE review is to identify hazards and prevent them from occurring or to mitigate their effects.
HF Skills Training	A series of training modules intended to provide a basic understanding of HF for the project team and to identify the HF issues in the Project.

Table 21-1 *(continued)*

HF Tool	Purpose
PHA review	The Process Hazard Analysis (PHA) review consists of a series of questions designed specifically for each of 13 pieces of equipment. Each question identifies the most important HF characteristics of the targeted equipment and /or system so the team can quickly decide whether a potential problem exists and what action to recommend (API, 2004).
Task Analysis	A tool that provides a structured, systematic approach to identify and analyze the tasks that are performed on the unit that are critical to the safe operation of the process. The purpose of a task analysis is to identify the human-machine interface issues that require solutions (Drury, 1983).
LINK Analysis	The objective of this tool is to optimize the location of people and equipment in a workplace. The purpose is to improve the work efficiency of operations and maintenance staff. (Sanders and McCormick, 1993).
Critical Procedure Development	To identify the critical tasks that operators are required to perform that could be affected by the design of equipment, then to determine the sequential activities that the operator must perform to successfully accomplish the task. For example, the design of filters can affect the methods used to change the cartridges.
QA/QC Review Process	Ongoing process that begins at the process specification phase and continues through to the end of detailed engineering. The process consists of training for the contract staff, completion of QA/QC checklists and reviewing the engineering design model to ensure that it complies with human factors design guidelines.
Pre-Startup HF Review	Purpose is to conduct a structured walkthrough through the project using a Human Factors Checklist and also to check the status of all HF follow-up items from the HF tracking database. The same review is conducted for MAJOR and LOCAL projects. The review is conducted at about 90% of plant completion.
HF Awareness for construction workers	Consists of awareness training sessions conducted by contractor specialty. The purpose of the training is two-fold. To build the project without injury to the contractor work force, and to ensure that the facility is fabricated to be safe and efficient for the owner to operate and maintain. Demonstrates the proper methods for field-run piping and wiring.
Post Project Review	The post project review has two objectives. First to capture those learnings that will help improve the project execution process. The second objective is to benchmark the performance of the project against other projects within the company and outside. The tool to accomplish the above objectives consists of a post-project review checklist. A portion of the PPR checklist, published in Attwood et al. (2003), that was designed to capture project learnings is given in Table 21-2

Examples of checklists that may be useful to a project team as they deal with ergonomic issues are provided in Tables 21-3 and 21-4.

A major integrated oil company found that the application of Human Factors during front-end engineering and design (FEED) reduced the number of design change requests during engineering design and construction by a factor of 10.

Table 21-2: Questions to capture Human Factors learnings from a project (partial table from Attwood et al. 2003)

Question	Response/Discussion
1. What HF issues were identified during each phase for the project and added to the HF Issues Tracking List?	
2. Did we identify the right things?	
3. Did we identify them at the right time?	
4. Did we use the proper tools and processes?	
5. Did we engage the right individuals in the process?	
6. Did we follow-up properly?	

Table 21-3: Human Factors Checklist for use when conducting a retrofit of an existing facility

1. Identify Jobs with Ergonomic Problem	Consult medical logs. Consult safety reports Consult with operating and maintainance personnel Conduct a pain and symptom survey
2. Evaluate and Investigate	Conduct a task analysis
3. Develop Alternative Solutions	Administrative solutions Engineering solutions Or, a combination of the two
4. Select Most Appropriate Solution	Base on consideration of funding, time, and effectiveness, etc.
5. Implement	Install selected solution
6. Follow Up and Audit	Determine effectiveness of selected solution If needed, develop and schedule as appropriate another solution

Table 21-4: Human Factors considerations in the design of equipment

1. Determine user requirements	List all the things users will be required to do.
	List all functions and tasks to be performed by people, including service and maintenance people.
2. Determine initial design specifications	Ask who is likely to be the user on this system.
	Assess the task considering the characteristics of the user population.
3. Determine final user design specifications	Consult an ergonomics design guide for dimension and sizing information.
	Specify design information (dimensions).
	Design for anticipated user population, not the current population.
4. Design/Specify equipment	Design actual equipment.
	Specify actual equipment.
5. Test new design/equipment	Where possible, make and test a mock-up with user population.
	Redesign as necessary.
6. Get feedback from initial users and redesign as appropriate	Get feedback on design as used in a production environment.
	Redesign if ergonomic goals are not met.
7. Continue to get feedback on design	Try to get information on the design throughout its life. In this way, mistakes can be avoided in future designs.

21.3 REFERENCES

API (2003) "Tool for incorporating Human Factors during process hazard analysis (PHA) Reviews of plant designs" November 11, 2003, (Washington, DC: American Petroleum Institute),.

Attwood, D. A., and Fennell, D. J. (2001) "Cost effective Human Factors Techniques for Process Safety," *CCPS International Conference and Workshop,* October 2–5, 2001, Toronto, ON, CANADA (NY: AICHE Center for Chemical Process Safety).

Attwood, D. A., Deeb, J. M., and Danz-Reece, M. E. (2003) "Ergonomic solutions for the process industries." [Burlington, MA: Gulf Professional Publishing (Elsevier)].

Drury, C.G. (1983) "Task Analysis methods in industry." *Applied Ergonomics, 14*(1), pp 19–28.

Robertson, N. (1999) "Starting Right: Sable Offshore Energy Project's HFE Program" *Proceedings, 1999 Offshore Technology Conference,* Houston TX, May 3-6 (Richardson, TX: Offshore Technology Conference).

Sanders, M. S. and McCormick, E. J. (1993) "Human Factors in Engineering and Design, Seventh Edition." McGraw-Hill, New York, NY.

CHAPTER 22

Procedures

22.1 INTRODUCTION

Procedures are a core part of every process operation. They are important because they provide rules to follow and standardized records of safe and approved operations and maintenance practices. They also provide consistent information across the plant and help minimize guesswork, leading to more efficient and safe operations. Certain aspects of procedures may be locally regulated. The U.S. OSHA 29CFR 1910.119 regulation on the Process Safety Management of Highly Hazardous Chemicals (OSHA, 1993) requires that all safety critical work must have written procedures.

There are four major types of procedures in process operations:

- *Start-up* and *shut-down procedures* are written instructions on how to bring the process up to operating conditions and how to bring the process down to a non-operating state. Both of these procedures apply to planned and not emergency situations.

- *Operating procedures* are written step-by-step instructions and associated information (cautions, warnings, notes etc.) for safely performing a task within operating limits.

- *Emergency operating procedures* are written instructions that address actions that must be taken to place a process in a safe and stable mode following a system upset. This could also include *emergency shutdown procedures*.

- *Maintenance procedures* are written instructions that address material control and practices needed to ensure system operability and integrity as well as maintenance, testing and inspection frequency.

Studies of accidents in a major petrochemical company (Attwood, 2005) show that 60% of incidents related to human performance were due to ineffective, incorrect, or missing procedures. Discussions with petrochemical plant operators indicate a variety of reasons why procedures are not followed. For example, the equipment required to perform the task (and specified by the procedure) is not available. In addition, the operating procedures:

- aren't available or readily accessible
- conflict with actual operations

- are outdated
- are confusing

There are also reasons that implicate the operators' behaviors. This includes: the operators were in a hurry, they were doing the job the "old" way, they were too experienced to refer to a procedure, they felt it was acceptable to deviate from a procedure, or they were unaware that a procedure exists. These are barriers to using procedures that must be identified and overcome to achieve compliance with procedures.

A raw material was undercharged to a reactor. As additional reactants were added (in correct amounts) later in the batch, there was not enough of the initial reactant available. This caused additional reactants to react with themselves creating the unexpected reaction. This resulted in rupture of the reactor and a fire. This trapped the operators for several minutes until the deluge systems extinguished the fire.

The undercharge was caused by:

1. Confusing procedural instructions which led to the raw material undercharge
2. Failure to cross-check amounts added to the reactor against the amount taken from storage tanks before adding remaining ingredients.

22.2   TOOLS

There should be a clearly defined process in place to identify jobs that need procedures in the field or for reference before going into the field. Haas (1999) recommends that the need for procedures is best determined by: (1) complexity of the tasks, (2) frequency with which the tasks are executed, and (3) the consequences of possible errors made while executing the job tasks. These needs are expanded in Table 22-1.

To develop effective procedures, the following process should be followed:

1. Identify and prioritize those jobs that require procedures.
2. Ensure that the procedure reflects the way the work is done and is technically accurate.
3. Ensure that procedures are written at the reading level of the users.
4. Ensure that the detail of procedures is not so limited that novice employees won't understand them, nor so detailed that experienced employees will not read them.

Table 22-1: Guidelines for when a procedure is required. Reproduced from Attwood et al. (2003). Based on Haas (1999).

Work Task Characteristic	Procedure needed in-hand, or checklist, or sign-off steps	Procedure available for reference, or reviewed prior to start	Procedure not needed but learned in training
Severity of consequence if error is made	Moderate to High (e.g., injury, process delay, equipment damage)	Moderate (some impact on process or safety)	Low (no impact on process or safety)
Complexity	Moderate to High (5 to 9 procedure steps, quick decisions)	Moderate	Simple
Frequency	Infrequent (less than once per month) to very frequent (weekly)	Infrequent to frequent	Frequent (multiple times per week)

5. Encourage procedure users to actively participate in the development and review of the procedures.

6. Develop processes to ensure that users are able to quickly and accurately locate the correct procedure for the job.

7. Write all procedures in a standard format which is set by a "style guide."

8. Instruct users in the use of their procedures.

9. Establish methods to ensure that procedures are reviewed and updated when the task, equipment or process changes.

10. Develop a system to ensure that shared knowledge between operations, maintenance and engineering is routinely incorporated into procedure updates.

11. Advise users when a procedure changes and instruct them on the changes.

12. Develop a system to learn from incidents, abnormal situations, near misses, behavior observations, and simulations and exercises and use the information to update procedures.

13. Use a checklist for critical procedures.

Proper design of procedures requires consideration of the following:

1. *Completeness and accuracy:* Does the procedure have enough information for the user to perform the task safely and correctly?

2. *Appropriate level of detail:* Has the level of detail considered the experience and capabilities of the users, their training and their responsibilities?

3. *Conciseness:* Conciseness demands eliminating detail and language that does not contribute to work performance, safety, or quality. Conciseness also means including only "need-to-know" and omitting "nice-to-know" information.

Table 22-2: Example procedure checklist (Attwood et al, 2003)

1. PROCEDURES EVALUATION CHECKLIST	

Procedures Checklist *(based on CCPS, 1996 and NUS, 1995)*

Procedure Title: _____ Plant site: _____

Date of evaluation: _____ Evaluator: _____

Checklist Item	Confirmed (✓)

Presentation and Usefulness
1. Does the procedure appear to be concise and easy to use?
   - No details that do not contribute to work performance, safety or quality
   - Only "need-to-know"
   - No "nice-to-know"
2. Is the procedure written so that the detail is appropriate to the range of experience of the users and their capabilities?
   - Technical terms familiar to the reader
   - No jargon
   - All acronyms are familiar
3. Is the procedure written so that the detail is appropriate to the complexity of the job including:
   - Criticality?
   - Potential hazards?
   - Ease of performing?
4. Are conditional instructions easy to understand? If an action must meet more than two requirements, are the requirements listed?
5. Are calculations clear and understandable? For complicated or critical calculations, is a formula or table included or referenced?
6. Can graphs, charts, and tables be easily and accurately extracted and interpreted?
7. Are steps written in short, concise, statements?
8. Are the same terms used consistently for the same components or operations?

Format, Layout, and Design
9. In general, the procedures:
   - Contain plenty of white space?
   - Contain tab markers to help locate them quickly, e.g. black tabs along the right edge of the page to help when searching through and locating major sections?
   - Are not cluttered or busy?
   - Use lines or white spaces to separate groups of related items?
   - Are written in **Times Roman?**
   - Use consistent font size throughout and at least 12-point?
   - CAPS are used for major titles?
   - Use mixed, upper and lower text throughout?
   - Left-justify text?
   - Identify steps by their own unique number
   - List each step in sequential order as it should be performed?
   - Begin each step with an action verb?

Table 22-2: *Continued*

Checklist Item	Confirmed (✓)

Format, Layout, and Design (*cont.*)

10. Are CAUTIONS, WARNINGS, and NOTES placed immediately before the step to which they apply?
11. Do CAUTIONS, WARNINGS, and NOTES stand out from the procedures steps?
12. Are P&IDs or flow charts placed ahead of the relevant steps or included with them?
13. If conditions or criteria are used to help the user make a decision or recognize a situation, do they preceed the action?
14. Is the procedure written for the lowest education level allowed among user (plant or contractor) personnel, e.g.
    - Low syllable count per word
    - Complexity count = 1 [complexity count is calculated by dividing the number of verbs in action steps by total number of steps in the procedure]

Content

15. Does the title accurately describe the nature of the activity?
16. If the procedure is over 5 pages, is it equipped with a Table of contents on the first page?
17. The first page of the procedure has the following information:
    - Procedure *Title*
    - *Objective* (purpose): Clearly defines the goal of the procedure
    - *Background:* Tells the user how this procedure fits into the "big picture" of the process and why it is important
    - *References:* Documents that support the use of the procedure
    - *Special equipment list:* Any special equipment that must be ready before starting the procedure
    - *Precautions* (Hazard summary): Conditions, practices or procedures that must be observed to avoid potential hazards
    - *Prerequisites:* Any initial conditions that the worker must satisfy or actions that need to be performed before starting the procedure
    - *Contact person or author*
        - *Authorized Signature*
18. Is the necessary procedure control information included on *each* page such as:
    - Facility or unit name or identifier
    - Procedure title
    - Procedure number
    - Date of issue, approval date, required review date and effective date
    - Revision number
    - Page number and total pages
19. Is the last page of the procedure clearly identified?
20. Are temporary procedures clearly identified?
21. Does every procedure have a unique and permanent identifier?
22. For duplicate processes, are the procedures complete and accurate for each process?

(*continued*)

Table 22-2: *Continued*

Checklist Item	Confirmed (✓)

Content (*cont.*)

23. Is all information necessary for performing the procedure included or referenced in the procedure?
24. Does the procedure include all steps required to complete a task? E.g. are any steps missing? (i.e. It does not reference other procedures to describe the step)
25. Does the procedure match the way the task is done in practice? E.g. any steps out of sequence?
26. Are all items referenced in the procedure listed in the "References" section of the procedure?
27. Are items listed in the references section of the procedure correctly and completely identified?
28. Do the references contain a list of supporting documents and locations?
29. If more than one person is required to perform the procedure, is the person responsible for performing each step identified?
30. Are steps which can be done simultaneously noted?
31. Is a signoff line provided for verifying critical steps of a procedure? (optional)
32. If the procedure requires coordination with others, does it contain a checklist, signoff, or other method for indicating the steps or actions have been performed or completed?
33. If a step contains more than two items, are they listed rather than buried in the text?
34. If two actions are included in a single step, can the actions actually be performed simultaneously or as a single action?
35. Are steps that must be performed in a fixed sequence identified as such?
36. Are operating or maintenance limits or specifications written in quantitative terms?
37. Does the procedure provide instructions for all reasonable contingencies?
38. If the contingency instructions are used, does the contingency statement precede the action statement?
39. Do procedures that specify alignment such as valve positions, pipe, and spool configurations, or hose station hook-ups:
    - Specify each item?
    - Identify each item with a unique number or designator?
    - Specify the position in which the item is to be placed?
    - Indicate where the user records the position if applicable?
40. Do emergency operating procedures contain provisions for verifying:
    - Conditions associated with an emergency? (initiating conditions)
    - Automatic actions associated with an emergency?
    - Performance of critical actions?
41. Do maintenance procedures include required follow-up actions or tests and tell the user who must be notified?
42. If a procedure must be performed by someone with a special qualification, are the required technical skill levels identified?

4. *Consistent presentation:* This element ensures that the procedure is readily comprehensible. It demands the use of:
   - A consistent terminology for naming components and operations.
   - A standard, effective format and page layout.
   - A vocabulary and sentence structure suitable for the intended user.
5. Administrative control: All procedures must be reviewed thoroughly before use and periodically thereafter.

The effectiveness of procedures is evaluated in a number of different ways, including (Attwood et al., 2003):

1. Observing multiple operators performing the procedure and comparing the way they perform the procedure to the documented procedure to identify deviations.
2. Evaluating the written procedure with a checklist. The checklist contained in Table 22-2 contains detailed questions that can be used to evaluate written procedures.
3. Conducting a hazard analysis.

22.3 REFERENCES

Attwood, D. A., Deeb, J. M and Danz-Reece (2003) "Ergonomic Solutions for the Process Industries." (Burlington, MA: Butterworth Professional Publications—Elsevier).

Attwood, D. A. (2005), personal communication.

Haas, P. M. (1999) "Human Performance Engineering: A Practical Approach to the Application of Human Factors." Paper No. NSC-99-108. *1999 NPRA National Safety Conference,* Dallas TX, (Washington, DC: National Petrochemical and Refiners Association).

OSHA (1992) "Process Safety Management of Highly Hazardous Chemicals" CFR 1910.119 (Washington, DC: U.S. Department of Labor, Occupational Health and Safety Administration).

22.4 ADDITIONAL REFERENCES

CCPS (1996) "Writing Effective Operating and Maintenance Procedures" (NY: AICHE Center for Chemical Process Safety).

NUS (1995) "Procedure Writing Workshop Manual," 7th ed. (Gaithersburg, MD: NUS Training Corporation).

CHAPTER 23

Maintenance

23.1 INTRODUCTION

Experience has shown that maintenance errors tend to fall into highly predictable and repetitive clusters. Therefore, these errors are relatively easy to identify and predict (see Figure 23-1). However, in many cases these errors remain undetected even if a functional test is carried out. In 1995 a Boeing 737 was forced to make an emergency landing on its first flight following a maintenance period. A non-destructive inspection of the engines had been carried out and the high pressure rotor drive cover on each engine had not been replaced. This resulted in a loss of almost all the oil from both engines.

Human factors has been recognized as a key issue in aviation maintenance and there is a wide range of reference material available both in hardcopy and on the internet (CAA, 2002). The aviation industry was the origin of reliability centered maintenance (RCM) (Lees, 2005). In RCM, equipment failure modes are identified, along with causes and consequences, in order to identify the equipment that requires higher maintenance.

So why is it so difficult to reduce maintenance error? One reason may be related to behavior. Many people find it difficult to accept that they make errors. Secondly, even if people do accept that they make errors, it will not necessarily prevent them from occurring. So why is maintenance error prone? In the maintenance environment, there are many situational and environmental factors that can degrade human performance and increase the likelihood of error. Examples of performance shaping factors (PSFs) are illustrated in Table 23-1.

23.2 IDENTIFYING CRITICAL MAINTENANCE TASKS

The first step in any assessment is to identify maintenance tasks involving critical safety equipment. Critical safety equipment could include relief devices, critical alarms, safety instrumented systems, etc. One approach would be through discussions with maintenance personnel. Alternatively, a more structured approach ranks the tasks with reference to critical safety items lists, analysis of previous records or the results of reliability, availability and maintainability (RAM) studies (Smith, 1997; Cox and Tait, 1998). The critical safety equipment can also be identified as the independent protection layers in layer of protection analysis (LOPA) (CCPS, 2001).

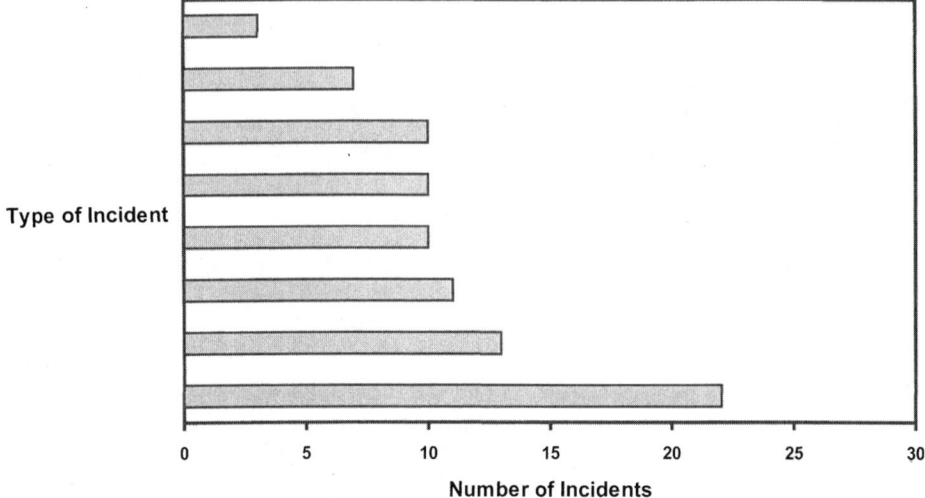

Figure 23-1: Typical Maintenance Errors (modified from Reason, 2003). The figure plots the number of incidents vs. the reason for maintenance failure.

23.3 MAINTENANCE PROTOCOL

One tool available for identifying human factors issues in maintenance is found in the HSE document "Improving Maintenance—a Guide to Reducing Error" (HSE, 2000). This protocol has two checklists that can be used for the assessment—reducing the need for expert HF input. The protocol can be used on its own or in conjunction with a more detailed error analysis. A more simplified approach considers only two types of errors that can be made during maintenance:

- correct removal but incorrect reassembly,
- omission of an action or part.

Often the typical errors are well-known by the maintenance staff—their involvement will highlight these errors in the analysis.

Table 23-1: Common performance shaping factors (PSF) affecting maintenance

Task difficulty, through poor design	Poor housekeeping/tool control
Poor environmental conditions	Lack of knowledge and experience.
Use of PPE	Time pressure
Poorly designed documentation/procedures	Poor communication and co-ordination
Lack of correct tools and equipment	Fatigue

A worker was told to replace a thermocouple in a high pressure process stream. The thermocouple was contained within a housing so that it could be replaced while the process was still operating. Unfortunately, the worker removed the housing, rather than the thermocouple, and process fluid escaped.

23.4 TOOLS

There are a number of methods to mitigate and reduce maintenance errors.

Task/Design Interventions

The whole point of maintenance is to remove, check and replace parts when considered necessary. This provides many opportunities for human error. It is possible, however, to design parts that can only be fitted in one way and to provide reminders, cues and other feedback to provide opportunities for error recovery.

Visibility and Signage

Clear signage and labeling of equipment reduces errors by preventing the maintenance staff from performing work on the wrong equipment. High visibility tags attract attention to equipment that is out-of-service and can ensure that this equipment cannot be brought back into operation without the required authorization and notification.

Maintenance Management

Maintenance tasks must have clearly assigned roles and responsibilities. Maintenance staff must be trained and competent in these defined responsibilities. Many organizations now use a maintenance resource management system to plan and manage maintenance activities. This is particularly useful during turnarounds where historically there is a higher rate of injuries to maintenance staff.

An important part of maintenance management is the layout design of the maintenance stores and placement of the maintenance supplies. It is critical that replacement in kind is done effectively. The chance for selecting the wrong gasket or welding rod, for example, must be minimized. Checks against the work order by maintenance stores personnel and the maintenance employee performing the work help avoid errors.

The design of I&E shops is also important from a human factors standpoint (see Chapter 8).

Procedures/Checklists

The quality of maintenance procedures is an important factor in managing maintenance errors. Procedures should be appropriate in length for the task and with consistent with the level of staff training (see Chapter 22 on Procedures). *Procedure evaluations* help to optimize operating procedures to increase usage and improve compliance.

A job safety analysis (JSA) of the procedure for that particular task on that particular system at that particular time can minimize the chance for errors.

Access for Maintenance

Sufficient forethought must be made in design for both routine and non-routine maintenance tasks. Important issues include: placement of work ladders and platforms; isolation of equipment; movement of heavy equipment; and access for vessel entries. Sufficient space should be provided so maintenance activities can be performed without placing personnel at risk from an ergonomic standpoint.

Maintenance by its nature frequently involves removing safeguards that are in place during normal operation. This includes handrails, machine guards, and so forth. Also, when equipment is removed, there is the potential to open up access to hazardous areas such as pits and sumps.

Safe Systems of Work

Due to the nature of maintenance, a safe system of work is required in order to perform maintenance tasks safely. Often the level of risk associated with maintenance tasks warrants the use of a permit to work (PTW) system. The PTW process should include: a risk assessment to identify the hazards associated with the task; the identification of the safeguards required, including any necessary isolations; and the checks and signatures that are required prior to, during and following the task. It is important to ensure robust procedures are in place for testing, hand back and for shift handovers.

Other Performance Shaping Factors

Shift scheduling and *fatigue assessments* help to reduce the physiological impacts on performance. (see Chapter 17 on Shiftwork Issues).

23.5 REFERENCES

CAA (2002), CAP 718: Human Factors in Aircraft Maintenance and Inspection, (London, Civil Aviation Authority), Available at: http://www.caa.co.uk.

CCPS (2001), "Layer of Protection Analysis—Simplified Risk Assessment" (NY: AICHE Center for Chemical Process Safety).

Cox, S. and Tait, R. (1998), "Safety, Reliability and Risk Management—An Integrated Approach," 2nd ed. (Amsterdam: Elsevier).

HSE (1991), "Developing Best Operating Procedures: A Guide to Designing Good Manuals and Job-aids" (London: U.K. Health and Safety Executive).

HSE (2000), "Improving Maintenance: A Guide to Reducing Human Error," (London: U.K. Health and Safety Executive).

HSE (2004), Research Report 213: "Human Factors Guidance for Selecting Appropriate Maintenance Strategies for Safety in the Offshore Oil and Gas Industry" (London: U.K. Health and Safety Executive).

Mannan, S. (2005), "Lees' Loss Prevention in the Process Industries," 3rd ed., Section 7.26: Reliability Centered Maintenance (Amsterdam: Elsevier).

Reason, J. and Hobbs, A (2003), "Managing Maintenance Error: A Practical Guide" (Aldershot, U.K.: Ashgate Publishing).

Smith, D. J. (1997), "Reliability, Maintainability and Risk—Practical Methods for Engineers," 5th ed. (Amsterdam: Elsevier).

CHAPTER 24

Safe Work Practices and Permit-to-Work System

24.1 INTRODUCTION

OSHA's Process Safety Management (PSM) Standard (OSHA, 1992) requires employers to develop and implement safe work practices (SWPs) to control hazards during operations such as lockout/tagout, confined space entry, opening process equipment or piping, and entrance into a facility by maintenance, contractor, laboratory, or other support personnel. SWPs often apply to hazardous work (HSE, 2002). For example, a permit is required for hot work by the PSM standard. The use of permits is governed by a set of procedures, called a permit-to-work (PTW) system. OSHA regulations exist for other types of hazardous work such as confined space entry (OSHA, 1998) and lockout/tagout (OSHA, 1996).

Many accidents have occurred during the performance of hazardous work (Kletz, 1988). Worker fatalities and injuries have occurred when confined spaces containing unsafe atmospheres were entered, piping containing hazardous chemicals was opened, and work was performed on equipment while still energized. Such accidents can often be attributed to human factors issues. A survey by the UK Health and Safety Executive found that the largest single cause of maintenance-related accidents was the lack of or deficiency in PTW systems (HSE, 2002).

A company considered hot work to be staightforward and maintained a supply of pre-signed permits for use by anyone conducting hot work. An untrained contractor used one of the permits to cut open a line containing flammable material causing a fire.

24.2 ISSUES

The following human factors issues should be addressed in designing, implementing and operating safe work practices and PTW systems.

People

Only authorized and qualified personnel should perform hazardous work. See Chapter 29 on Competence Management for more information. Hazardous work, by its nature, may be more stressful than normal work. Consequently, human factors such as stress, fatigue and shiftwork are particularly important. See Chapter 17 on Shiftwork Issues for more information. Other individuals involved in hot work such as the permit issuer, fire watch, and rescue personnel should also be qualified.

Training and Briefing

Involved personnel should be appropriately trained and briefed before they perform work. The training should be relevant to the actual work and circumstances under which it will be performed. Persons performing hot work should be briefed on the work to be performed at the job site so that the equipment and hazards involved and the precautions required can be pointed out. Personnel performing hazardous work often use specialized equipment such as hoists and other lifting equipment; personal protective equipment (PPE) such as supplied-air systems; and safety equipment such as rescue lines. Training should include the use of such equipment. Personnel should also be trained in the recognition of when SWPs and a permit are needed.

If hazardous work is performed infrequently, the need for refresher training is particularly important. The nature of the work to be performed can vary from job to job so that specific briefings may be needed. Affected personnel in operations, maintenance, safety, and others such as contractors and temporary employees, should be included, as appropriate. See Chapter 12 on Training for more information.

Communications

PTW systems are in large part communications systems for facility personnel. Personnel performing hazardous work need to communicate with each other, standby personnel, operating personnel, and supervisors.

Communications may be in written form such as procedures, permits, labels, and signs; it may be verbal such as spoken instructions; or it may be nonverbal such as hand signals in the field. In some situations two forms of communication should be used such as both verbal and written instructions. The means of communicating critical information should be specified.

Hazardous work should be properly described for workers by responsible personnel. Figures and drawings may be useful to assist in describing the work, its location and limitations. Those who perform hazardous work should be given the opportunity to discuss the work with those authorizing it. It is poor practice to leave a permit on a desk to be picked up because it discourages communication.

Permits should be suitably displayed, e.g. in the permit coordination room, main control room, local control rooms, and at the work site. Entries in permits should be legible and permits should be able to withstand the environments in which they will be displayed, such as at the work site.

Permits should be cross-referenced where there is interaction between jobs, including isolations if they are common to more than one job. Particular attention should be paid to the hand back from personnel performing hazardous work to those responsible for its operation.

See Chapter 13 on Communications for more information.

Procedures

Hazardous work should be performed according to established safe work practices. The procedures should fully document how the SWPs and PTW system works, the circumstances under which it should be used, and the responsibilities of those involved. See Chapter 22 on Procedures for more information.

Precautions

Personnel carrying out hazardous work and process operators must be aware of, understand, and carry out those actions needed to ensure the work is performed safely and does not adversely impact people or the process. Permit issuers should be sufficiently knowledgeable and able to determine the hazards involved and the precautions needed for hazardous work. Permits should designate individuals for special roles such as fire watch and rescue. Isolation requirements should be specified.

Precautions must be specified, communicated, and understood. PTW systems should verify that these requirements have been met.

Identification

Significant potential exists in hazardous work for the wrong piece of equipment or the wrong location to be selected. Correct and clear identification of equipment and locations is critical. This should be covered in writing in the permit and reinforced verbally, ideally in the field. Chapter 11 on Labeling provides more information relating to identification.

Checks

Critical actions in performing hazardous work, such as confirming isolations are in place, should be checked independently by someone other than the individual performing the action. Confirmation should usually be documented by signature.

Environmental Factors

Hazardous work is often conducted in adverse working environments. Access and working space may be restricted. Physical movement of workers may be impeded by the need to wear PPE. Environments may be hot, humid, cold, noisy, and so on. The potential impact of these factors on hazardous work should be addressed. See Chapter 15 on Environmental Factors for more information.

Equipment and Process Design

The usability of tools and equipment employed in performing hazardous work and the way in which personnel interact with the process while carrying out work affect the likelihood of human failures. Such failures are particularly significant since they occur during the performance of inherently hazardous activities and may have significant impacts on the workers involved. Specialized tools and equipment are often used in hazardous work, for example, welding equipment, breathing equipment, and lifting equipment, and workers may be less familiar with them and less skilled in their use. Necessary tools, equipment, parts, supplies and materials should be specified and their availability ensured. See Chapter 8 on Facilities and Workstation Design and Chapter 4 on Process Equipment Design for more information.

Unauthorized Changes

Deviations from SWPs, changes in materials, tools, or equipment, and so on should not be permitted without a suitable review. A separate permit should be issued if it is determined that other work is needed during the performance of permitted work. See Chapter 25 on Management of Change for more information.

Time Constraints and Pressures

In some circumstances there may be time pressures on performing hazardous work that may result in cutting corners and short cuts. For example, excessive workloads may occur during turnarounds, revamps, etc. and PTW systems can be overloaded. PTW systems should be designed to accommodate such situations, for example, by ensuring a sufficient number of trained contractor personnel will be available and that the PTW system does not include non-critical activities. See Chapter 16 on Workload for more information.

Issuing permits with a duration longer than the job is reasonably expected to take should be avoided. Generally, multiple-day permits should not be issued since conditions can change from day to day, even with unknowns such as the weather. A renewed permit should be issued each day.

Hazardous work should be properly described and authorized by a responsible person. Responsibilities for hazardous work, including safety precautions, should be defined and assigned. Personnel responsible for overall control of the work and its execution should be specified. Needed resources should be available. Appropriate supervision of personnel using the PTW system should be provided and SWPs and the PTW system should be properly managed, inspected and reviewed.

Culture

Some personnel may resist the use of SWPs and PTW systems. They may not be familiar with how these systems work or believe they are not needed. Other personnel may embrace them as a significant contribution to safety. See Chapter 19 on Safety Culture for more information.

Permit Design

Permits should be tailored to the type of hazardous work so that emphasis is given to the particular hazards present and the precautions required, i.e. separate permits should be used for different types of work. Standard formats should be used to facilitate training and reduce the chance of failures. Permits should be readily identifiable, clearly laid out, legible and readily usable under the circumstances in which they may be employed. They should contain document control information including a revision number and issue date. Statements or questions that could be ambiguous or misleading should be avoided. See Chapter 14 on Documentation Design and Use for more information.

PTW System Design

The PTW system should match the organization's structure and procedures. PTW systems should not be used unless they are really needed since overuse will diminish their value.

> An explosion in the gas compression module of the Piper Alpha oil production platform in the North Sea resulted in 167 fatalities. A key contributing cause of the accident was the lack of an effective PTW system. This resulted in condensate being admitted to a pump from which the pressure safety valve had been removed. An atmospheric release occurred followed by an explosion and fire.

24.3 TOOLS

The human factors issues described above can be used as design criteria for SWPs and PTW systems. Guidance on the design and use of SWPs and PTW systems is available in the literature (HSE, 1997; HSE, 2002; Lees, 1996; CCPS, 1995).

Hazard analysis and risk assessment methods can be used to identify human failures and human factors issues for hazardous work. The analyses identify possible errors and the factors that influence their likelihood. See Chapter 26 on Qualitative Hazard Analysis, and Chapter 27 on Quantitative Risk Analysis for more information.

Regular monitoring is needed to ensure correct implementation and use of SWPs and PTW systems. Spot checks on permits should be performed and comparisons made of permits with the work being performed. Deviations should result in immediate corrective action. Audits against best-practices using checklists should be conducted at defined intervals (CCPS, 1993). Reviews of the system effectiveness should be made at longer intervals with input sought from involved workers and supervisors.

24.4 REFERENCES

CCPS (1993), "Guidelines for Auditing Process Safety Management Systems" (NY: AICHE Center for Chemical Process Safety).

CCPS (1995), "Guidelines for Safe Process Operations and Maintenance" (NY: AICHE Center for Chemical Process Safety).

HSE (1997), "Guidance on Permit-to-Work Systems in the Petroleum Industry" (Sudbury, UK: HSE Books).

HSE (2002), "Permit-to-work systems" (Sudbury, UK: HSE Books).

Kletz, T. A. (1988), "What Went Wrong? Case Histories of Process Plant Disasters", 2nd ed. (Houston, TX: Gulf, Houston, TX).

Lees, F. P. (1996), "Loss Prevention in the Process Industries. Hazard Identification, Assessment and Control," 2nd ed. (Oxford, UK: Butterworth-Heinemann).

OSHA (1992), "Process Safety Management of Highly Hazardous Chemicals; Explosives and Blasting Agents," 29 CFR 1910.119 (Washington, DC: Occupational Safety and Health Administration), published 2/24/1992 and effective 5/26/92.

OSHA (1996), "The Control of Hazardous Energy (Lockout/Tagout)," 20 CFR 1910.147 (Washington, DC: Occupational Safety and Health Administration).

OSHA (1998), "Permit Required Confined Spaces," 20 CFR 1910.146 (Washington, DC: Occupational Safety and Health Administration).

24.5 ADDITIONAL REFERENCE

HSE (1997), "The Safe Isolation of Plant and Equipment" (Sudbury, UK: HSE Books).

CHAPTER 25

Management of Change

25.1 INTRODUCTION

Human factors considerations enter into management of change in several ways, including:

- Process changes may affect the way people interact with the process. For example, changes in chemicals, technology, equipment, procedures, or facilities may impact process safety by increasing the likelihood of human failures when using the chemicals, technology, equipment, procedures or facilities. Such human factors issues should be addressed when assessing the impact of proposed changes.

- Changes related to people and organizations may have implications for process safety. This includes changes in personnel assignments, roles and responsibilities, reporting relationships, abilities and competencies, organizational structure, management systems, goals and objectives, equipment operation or use, manning levels and workloads, levels of automation, engineering and technical support, available resources, communications systems, etc. Many plants experience reorganization in the form of downsizing, re-engineering, centralization, mergers and acquisitions, etc. Regardless of the cause, organizational changes can cause temporary or permanent impacts on personnel and processes and possible adverse impacts on safety.

    Sometimes, the potential importance of such changes is not recognized. OSHA's Process Safety Management (PSM) standard, 29 CFR 1910.119 (OSHA, 1992) requires that companies address changes in process chemicals, technology, equipment, procedures, and facilities that affect a covered process. However, organizational and some other types of changes are not specifically addressed in the PSM standard.

- The process of managing changes involves a variety of human factors issues, including:

    - The willingness of personnel to use the Management of Change (MOC) procedure, because:
        - It may be viewed as a bureaucratic and paperwork-intensive requirement that interferes with production.
        - The plant may be operated differently between shifts.

- ○ There may be time pressures to make a change.
- ○ People may consider themselves to be competent to design and implement changes without the need for a formal review process.
- ○ People may consider a change to be minor and not worth consideration.
- ○ Consultation with the necessary people may be difficult.
- ■ Ability of personnel to properly use the MOC procedure:
  - ○ Personnel may not be aware of the procedure.
  - ○ Personnel may lack understanding of change types and review requirements.
  - ○ Signoffs may be inadequate and documentation incomplete.
- ■ Adequacy of training in MOC:
  - ○ Personnel may fail in their responsibilities owing to inadequate training.
- ■ Ability and willingness to properly review the change:
  - ○ Reviewers may be biased or have preconceived ideas.
  - ○ Time pressures may produce inadequate reviews.
- ■ Acceptance of change by affected personnel:
  - ○ People generally dislike change and may resist it.
  - ○ If people are not involved, they will likely not be committed. Thus, changes may not be accepted in practice by personnel.
- • The process of *implementing* changes can involve various human factors issues. Problems that may arise include:
  - ■ Improper implementation of the change (e.g. not installing as designed).
  - ■ Inadequate checking and verification that the change has been correctly implemented.
  - ■ Inadequate management of temporary changes (e.g. changes approved for a specific time period are extended without further review).
  - ■ Inadequate training of personnel on how the change affects their work.

Considerable time may be required to take a change from conception to completion and the MOC process often involves circulation of documentation for review and approval by various people. There is the potential for shortcuts and sloppiness and there are many human error pitfalls along the way. MOC training that provides examples and case histories of how things have gone wrong in making process changes is of considerable value in convincing people of the need to properly address MOC.

Approaches for handling these issues are not addressed in standard references on managing process changes (CCPS, 1992; CMA, 1993; Sanders, 1993; Sutton, 1998) other than brief mentions of the need to address organizational changes.

Figure 25-1 shows the damage resulting from the Flixborough, UK accident in 1974. This accident was the result of inadequate management of change.

Figure 25-1: The consequences of a vapor cloud explosion owing to inadequate change management. [Reprinted from HSE (2003).]

An engineer who had been assigned responsibilities for various PSM activities was laid off during company downsizing. However, his responsibilities were not reassigned. Work on his PSM activities ceased but the problem was not discovered until a PSM audit was conducted two years later.

25.2 TOOLS

The Chemical Manufacturers Association (now the American Chemistry Council) has published guidance for managing safety and health during organizational change (CMA, 1998). It provides sample worksheets and checklists. Human factors are not specifically addressed in this guidance but the checklists include some questions relating to human factors. The UK Health and Safety Executive (HSE) has also published guidance on managing organizational change (HSE, 2003). It deals with common pitfalls and suggests a framework for managing organizational change that includes consideration of human factors. The HSE and the UK Energy Institute have published guidance on de-manning (HSE/EI, 2005).

There are no other tools currently available that have been specifically developed to address human factors in managing process changes. However, tools identi-

Date:	Entry Author		Status:	MOC Number:

* Indicates Field must be filled in before saving document.

Title*

Area*   **Department***   **Equipment**

Plant General

Initiator, if Not Author Duration of Process Change*

MICHELLE BROWN   **Permanent or Temporary?**   Permanent Change?

                                              Temporary Change?

AFE / Work Order #   **Select if Off-shift:**   Off-shift Change ?

Type of Change

Are there new or updated SOPs associated with this MOC?*
(If you click Yes, revisions to the Operating Procedures will be required. You must enter the name responsible person in Section B before submitting this MOC to the PS&H Manager.)

Attach new or red-lined electronic SOP here: _How to Attach ??_

Enter a Detailed Description of the Proposed Change*:
Complete and specific description of WHAT is to be changed

Technical Basis of Proposed Change*:
Technical explanation of WHY the change is needed and how it will be implemented

Is there Paper (non-electronic) Documentation Associated with this MOC?

Mark the PSI to be Updated as a Result of this Change.

Specify the PSI that will Require Revisions by clicking Yes to that PSI.

Figure 25-2: An on-line management of change system.

Operations/Process :

	Process Safety Mgmt PSI	Revision Required		Process Safety Mgmt PSI	Revision Required		Process Safety Mgmt PSI	Revision Required
	Operating Manual			**Control System (cont'd)**			**Drawings**	
1.	Operating Procedures		10.	Alarm Limits		19.	P&IDs	
2.	Batch Sheets		11.	Logic Blocks		20.	PFDs	
3.	Process Description/Overview		12.	TDC Range Blocks			**General**	
4.	Safe Upper & Lower Limits for critical operating parameters			**Process Technology**		21.	On-Shift Training	
5.	Consequence(s) of Deviation from Standard Operating Conditions (SOC)		13.	Process Chemistry		22.	Off-Shift Training	
6.	Abnormal Operations		14.	Maximum Intended Inventory			Training Level	
	Process Control Documentation		15.	Physical Data			**Other Technical Information**	
7.	Control System Code		16.	Chemical Compatibility Matrix		23.	Other:	
8.	Control System Graphics		17.	Material Balance				
9.	Control System Configuration		18.	Energy Balance				

Maintenance :

	Process Safety Mgmt PSI	Revision Required		Process Safety Mgmt PSI	Revision Required		Process Safety Mgmt PSI	Revision Required
	Mechanical Integrity Documentation		30.	Specific Equipment Plan (SEP), (e.g., MI Updates)			**General**	
24.	Materials of Construction		31.	Suppression System		37.	Plot Plans	
25.	Electrical Area Classification		32.	Instrument/Electrical Loop Drawings		38.	On-Shift Training	
26.	Relief System Design		33.	Instrument Calibration Data Sheets		39.	Off-Shift Training	
27.	Ventilation System Design		34.	Instrument Data Sheets			Training Level	
28.	Interlock Data Sheets		35.	One Line Drawings (i.e., Power Distribution)			**Other Mechanical Information**	
29.	Detection Systems		36.	Equipment Specifications		40.	Other:	

Figure 25-2: *Continued*

Responsible Care :

	Process Safety Mgmt PSI	Revision Required		Process Safety Mgmt PSI	Revision Required		Process Safety Mgmt PSI	Revision Required
	MSDS Book			**General**			**Environmental**	
41.	Toxicity Information		48.	Emergency Plans		52.	Air & Operating Permits	
42.	Permissible Exposure Limits		49.	PHA & PSRP Documentation		53.	RCRA Permits	
43.	Physical Data		50.	Training		54.	Wastewater Discharge Permits	
44.	Reactivity Data			**Other Responsible Care Information**		55.	Risk Management Plan	
45.	Corrosivity Data		51.	Other:			**Industrial Hygiene**	
46.	Thermal and Chemical Data					56.	Personnel Protective Equipment Requirements	
47.	MSDS Form							

Action items may be entered, but SAR's cannot be created until an MOC number is assigned. To expand the PSI Action Item table for item entry, Click button:

PSI Action Items

SAR Assigned	Action Items	Priority	Responsibility	PPR Area	PPR Department	Target Date	Completion/ Cancel Date	Issued By

Figure 25-2: *Continued*

Checklist to be filled in by Initiator and reviewed by Process Safety & Health Manager

**Answer <u>All</u> Questions. Note any Follow up Issues in the Comments Column.
(Each "Yes" answer requires a Comment.)**

Process Safety & Health Manager :
Process Safety Code Specialist :

	Yes or No?	Comments
Part 1:		
Equipment/Instrumentation/Piping: *Will this change affect:*		
Service? (temperature, pressure, chemical, etc.)		
Failure mode? (e.g., open/close, hi/lo, etc.)		
Materials of construction?		
Utilities? (directly connected to the process)		
Size/capacity/performance/calibration?		
Equipment code certifications?		
Mechanical Integrity lists?		
Inerting, grounding, or bonding?		
*Another process or production area		
Decomissioning (Reference to MI Program)		
Procedures: *Will this change affect:*		
New or modified Operating Limits? (temp, pressure, flow, level, etc.)		
Clean-out/Decontamination Procedures?		
Emergency Response Procedures?		
Emergency Shutdown Procedures?		
New or Significantly modified Operating Procedures?		
*New Product for OSW		
Raw Materials/Work in Process/Final Products: *Will this change affect:*		
Composition or volume?		
Increase Static Potential?		
Decrease Flash Point?		
Increase Toxicity?		
Increase Corrosivity?		
Increase Personnel Exposure?		
Atmospheric Emissions?		
Liquid or Solid Effluent?		
*Product Quality? (Impurity profile or physical properties)		

Figure 25-2: *Continued*

*cGMP Process?	
Process Controls: **Will this change affect:**	
Process Interlocks?	
New or Modified Application Program?	
Emergency Shutdown Sequence?	
Critical alarm limits/priorities?	
Part 2:	
Responsible Care (Permits, Regulatory, etc.): **Will this change affect or involve:**	
Temporary Pilot Operations?	
Safety or Environmental Interlock systems? (including by-passing)	
Exothermic/Decomposition Potential?	
Performance of a Process Safety Device or System? (Relief valves, rupture discs, fire suppression systems, process vent systems, containment systems, explosion panels, etc.)	
Current Regulatory Permits? (EPA, RCRA, NPDES, CAAA, etc.)	
RMP Status or Application?	
Electrical Classification?	

When the PS&H Manager has reviewed Sections A, B & C, and completed Section D, Click button :
(Note: The Classification & Hazard Identification Checklist must be reviewed by the PS&H Manager before this MOC may be Submitted for Approval.)

Reviewed by:	**Date Reviewed:**

Figure 25-2: *Continued*

fied in other sections of this book that address human factors for equipment, procedures, process control, competence management, operator workload, etc. can be adapted to address human factors issues in managing changes. CCPS is currently preparing technical guidelines on management of change.

Figure 25-2 shows an on-line management of change form.

25.3 REFERENCES

CCPS (1992), "Plant Guidelines for Technical Management of Chemical Process Safety" (New York: AICHE Center for Chemical Process Safety).

CMA (1993), "A Manager's Guide to Implementing and Improving Management of Change Systems" (Washington, DC: Chemical Manufacturers Association, now the American Chemical Council).

CMA (1998), "Management of Safety and Health During Organizational Change" (Washington, DC: Chemical Manufacturers Association, now American Chemical Council).

HSE (2003), Chemical Information Sheet No. CHIS7: "Organisational Change and Major Accident Hazards" (London: U.K. Health and Safety Executive).

HSE/EI (2005), http://www.energyinst.org.uk/humanfactors.

OSHA (1992), "OSHA 1910.119: Process Safety Management of Highly Hazardous Chemicals" (Washington, DC: U.S. Occupational Safety and Health Organization).

Sanders, R. E. (1993), "Management of Change in Chemical Plants" (London: Butterworth-Heineman).

Sutton, I. S. (1998), "Management of Change" (Houston, TX: Southwestern Books).

CHAPTER 26

Qualitative Hazard Analysis

26.1 INTRODUCTION

Hazard analysis identifies possible hazard scenarios for a process or facility (CCPS, 1992). Qualitative risk estimates are often included in hazard analysis. OSHA's Process Safety Management (PSM) standard, 29 CFR 1910.119 and EPA's Risk Management Program (RMP) rule, 40 CFR Part 68, require that a process hazard analysis (PHA) be performed for processes covered by the regulations and that, among other things, "the PHA shall address human factors."

Hazard analysis must address human failures together with equipment failures and external events as possible causes of hazard scenarios (see Figure 26-1). Note that equipment failures may be attributed ultimately to human failures on the part of process designers, specification engineers, fabricators, maintenance personnel, etc. and some external events are also human induced. Human failures may be initiating events, intermediate events, or enabling events for hazard scenarios. Closing the wrong valve may be an initiating event, failure to respond to an alarm may be an intermediate event, and bypassing a trip may be an enabling event.

Hazard analysis should also address the human factors that influence the likelihood of human failures occurring. For example, an operator may fail to close a manual valve in a line when required to do so. This could be the initiating event for a hazard scenario. The human factors that influence the likelihood of this failure must be identified in order to 1) assess the likelihood of the failure, 2) identify existing safeguards that may protect against it, and 3) decide what recommendations may be needed to reduce the risk to a tolerable level. For example, the valve may not be labeled, it may be located close to another similar valve in an adjacent line, operator training or procedures may be inadequate, etc.

Treatment of human factors in hazard analysis requires:

- Identification of human failures as causes of or contributors to hazard scenarios.
- Identification of human factors that influence the likelihood of human failures.
- Optionally, qualitative assessment of the likelihood of human failures.

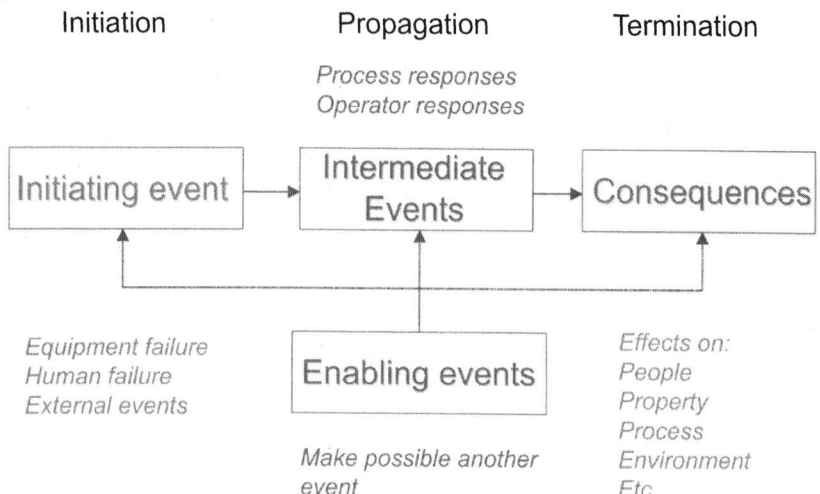

Figure 26-1: Elements of a hazard scenario. Copyright © 2004, Primatech Inc. All rights Reserved.

26.2   TOOLS—HUMAN FAILURES

Human failures are usually identified in hazard analysis by brainstorming causes of scenarios along with equipment failures and external events. Knowledge of human involvement in a process and the types of failures people may make is used to identify human failures. Structured brainstorming uses lists of the people involved in the process, the actions they perform, and the types of failures that may occur (acts of omission and commission, extraneous acts, and violations or deliberate acts) and combines entries from the three lists to identify possible human failures. A key to success in identifying human failures is insuring that all people involved in the process are considered, including operators, mechanics, contractors, and others. Human failures identified by hazard analysis can be seen in Figure 26-2 in the "Cause" column of the worksheet.

In hazard analysis, the likelihood of human failure is usually assessed qualitatively along with the likelihood of equipment failures and external events to provide an estimate of the overall scenario likelihood. Estimates are made using engineering judgement.

Also important in hazard analysis are human factors issues that affect the performance of studies. These include the willingness and ability of the analysts to consider human failures and human factors, the team's ability to work together effectively, and time pressures to complete the study.

Tools are also available to analyze the work performed by people to identify the failures that may occur. Task analysis is one such method (Kirwan and Ainsworth, 1992).

NODE: (1) LINE FROM ROAD TANKER TO STORAGE TANK					
PARAMETER: Flow			INTENTION: 100 - 200 GPM		
DEVIATION	CAUSES	CONSEQUENCES	SAFEGUARDS	RECOMMENDATIONS	E
No Flow	1. Unloading hose ruptured owing to truck movement (HF - truck is not properly spotted and chocked by a substitute driver without experience or training)	1.1. Spill of flammable material and possible fire	1.1.1. Spotting and chocking procedure 1.1.2. Flammable gas detectors	1.1.1. Consider adding to the tank truck unloading procedure an independent check on spotting and chocking the truck. 1.1.2. Qualify and screen truck drivers.	
	2. Unloading hose not connected to unloading line (HF - operator is distracted)	2.1. Same As 1.1	2.1.1. Same As 1.1.1 and 1.1.2	2.1.1. Consider adding to the tank truck unloading procedure an independent check on hose connections.	
	3. Unloading hose connected to wrong unloading line (HF - lines not labeled)	3.1. Inadvertent mixing of materials and possible reactivity incident	3.1.1. Operator training	3.1.1. Address this issue in the operator training program. 3.1.2. Label unloading lines. 3.1.3. Consider segregating unloading stations.	
		3.2. Possible overfilling of other tank	3.2.1. Tank high level alarm	3.2.1. None needed.	

Figure 26-2: Example of PHA with human failures and human factors identified.

A company performed a PHA on a process and completed a PHA report that was provided to management. However, the report was simply filed and no action was taken on the recommendations it contained until nearly five years later when a PHA revalidation was performed. The company took immediate steps to correct the problems identified and also changed the management system to prevent this from occurring in the future. The company considered itself fortunate that an accident did not occur as a result of the neglected recommendations.

26.3 TOOLS—HUMAN FACTORS

Human factors can be considered during hazard analysis (Bridges, et al., 1994). Simple checklists can be used to identify human factors issues that may impact human failures identified in the hazard analysis. Recommendations to deal with the human failures address the underlying human factors issues that make them likely.

An example of a checklist is provided in Figure 26-3. The use of checklists to screen for human factors during Hazard and Operability (HAZOP) studies has been described (Attwood, et al., 2004). The American Petroleum Institute (API) has developed human factors checklists for PHAs of new plant designs (API, 2003). The report states that the checklists are not intended to be applied retroactively. The questions include example situations and potential solutions to increase their utility.

Human factors can also be treated as adjunct studies to PHAs using checklists of typical issues for processes. An example checklist is provided in CCPS (2001). Checklists contain questions such as "Are valves labeled?" and "Are manual valves readily accessible to operators?" An example of part of a completed checklist is provided in Figure 26-3.

Approaches have also been developed to address human factors using specific hazard analysis methods. For example, some users of the HAZOP method include a human factors parameter and use it to identify and analyze human factors issues. An example is provided in Figure 26-4. Hazard analysis methods can also be applied directly to procedures to identify human failures and the factors influencing them (Bridges, et al., 1996). A modified Layers of Protection Analysis (LOPA) can be

TASK: (1) TANK TRUCK UNLOADING			
CATEGORY: (1) ACCESSIBILITY / AVAILABILITY OF CONTROLS AND EQUIPMENT			
QUESTION	A	REMARKS	RECOMMENDATIONS
1. Are adequate supplies of protective gear readily available for routine and emergency use?	Y	Fire extinguisher is located at unloading station.	None.
2. Are workers able to perform both routine and remergency tasks safely while wearing protective equipment?	NA		
3. Is emergency equipment accessible without presenting further hazards to personnel?	N	Fire extinguisher is located too close to source of fire in the event of a spill	Relocate fire extinguisher.
4. Is communications equipment adequate and easily accessible?	P	Operator has a radio but the truck driver, who may be left unsupervised, does not.	Consider providing a radio to the truck driver on arrival at site. Provide for sign out.
5. Is needed equipment readily accessible?	N	Chocks are sometimes misplaced.	Consider providing storage cabinet with lock at unloading station. Key to be kept by gate guard and signed out.

Figure 26-3: Example of a human factors checklist study.

NODE: (1) LINE FROM ROAD TANKER TO STORAGE TANK					
PARAMETER: Human Factors					
DEVIATION	CAUSES	CONSEQUENCES	SAFEGUARDS	RECOMMENDATIONS	B
Part Of Human Factors	1. Driver may be a substitute who is not experienced or trained	1.1. Truck is not properly spotted and/or chocked. Truck may move and unloading hose may rupture with flammable spill and possible fire	1.1.1. Spotting and chocking procedure 1.1.2. Flammable gas detectors	1.1.1. Qualify and screen truck drivers.	
	2. Operators are responsible for multiple unloading operations that may occur simultaneously in different locations	2.1. Unloading hose may not be connected to unloading line and flammable material may be spilled with a possible fire	2.1.1. Same As 1.1.1 and 1.1.2	2.1.1. Consider reallocating operator unloading responsibilities.	
	3. Unloading lines are not labeled	3.1. Unloading line may be connected to the wrong unloading line possibly causing a reactivity incident	3.1.1. Operator training	3.1.1. Label unloading lines. 3.1.2. Consider segregating unloading stations.	
		3.2. Wrong connection with possible tank overflow	3.2.1. Tank high level alarm		

Figure 26-4: Example of human factors treated using deviations in a HAZOP study.

used to identify the human factors involved for critical scenarios involving human failures (Baybutt, 2002).

A key to success is identifying human factors issues both locally within a process and also globally for the entire process. For example, readability of a particular gauge is a local issue while the quality of written operating procedures is a global issue.

26.4 REFERENCES

API (2003), "Tool for Incorporating Human Factors During Process Hazard Analysis Reviews of Plant Designs" (Washington, DC: American Petroleum Institute).

Attwood, D. A., Deeb, J. M., and Danz-Reece, M. E. (2004), "Ergonomic Solutions for the Process Industries" (Amsterdam: Elsevier), p. 407.

Baybutt, P. (2002), "Layers of Protection Analysis for Human Factors (LOPA-HF)", *Process Safety Progress,* Vol. 21, No. 2, pp. 119–129.

Bridges, W. G., Kirkman, J. Q., and Lorenzo, D. K. (1994), "Include Human Errors in Process Hazard Analyses," *Chem. Eng. Prog.,* March, Vol. 90, No. 5, pp. 74–82.

CCPS (1992), "Guidelines for Hazard Evaluation Procedures" (NY: AICHE Center for Chemical Process Safety).

CCPS (2001), "Revalidating Process Hazard Analyses" (NY: AICHE Center for Chemical Process Safety), Appendix G.

Kirwan, B. and Ainsworth, L. K. (1992), "A Guide to Task Analysis" (Taylor and Francis, London).

26.5 ADDITIONAL REFERENCES

API 770 (2001), "A Managers Guide to Reducing Human Errors; Improving Human Performance in the Process Industries" (Washington, DC: American Petroleum Institute). Note: This was previously published as CMA (1990), "A Manager's Guide to Reducing Human Errors" (Washington, DC: Chemical Manufacturers Association).

CCPS (1994), "Guidelines for Preventing Human Error in Process Safety" (NY: Center for Chemical Process Safety).

CHAPTER 27

Quantitative Risk Analysis

27.1 INTRODUCTION

In quantitative risk analysis, numerical estimates of the consequences and frequencies of hazard scenarios are used to develop estimates of risk for the process/facility under study (CCPS, 2000). Hazard scenario frequency is estimated by combining the frequency of the initiating event with the probabilities of other events that make up the scenario. Data on equipment failure rates, external event frequencies and human failure rates are needed to perform these calculations. Those aspects of the calculation that deal with human failures are often termed *human reliability analysis* (HRA). HRA provides quantitative values for human error rates for inclusion in system modeling techniques such as fault tree and event tree analysis. It considers the conditions that cause people to fail.

HRA involves:

1. Identification of tasks performed by plant personnel.
2. Task analysis to identify potential human failures.
3. Identification of conditions that affect human failure rates. These are called *performance shaping factors* (PSFs). They include such human factors as training, environmental conditions that may cause stress, effectiveness of training, readability of displays and controls, etc.
4. Application of data and/or expert opinion on human failure rates and PSFs to determine human failure rates and probabilities.

A critical check valve was installed in reverse by a mechanic and a hazardous material release occurred. The installation procedure required independent checks of the installation by both the maintenance supervisor and a second mechanic. However, each assumed the other had checked and that one check was sufficient. The assumed error rate for each check was 1 in 100 giving an acceptable overall failure rate of 1 in 10,000. However, the actual error rate for failing to check the installation was 1 since no check was performed!

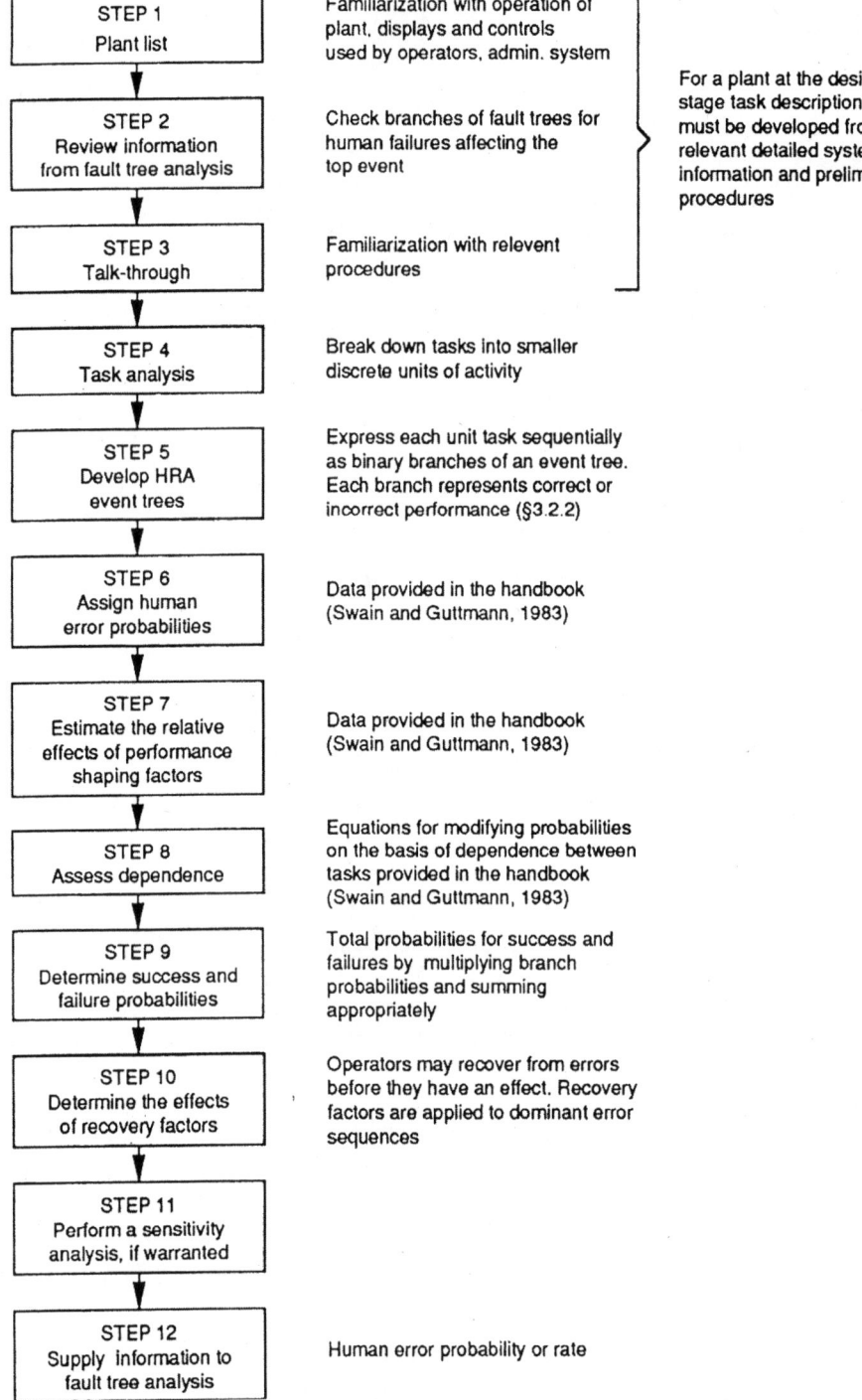

STEP 1 Plant list	Familiarization with operation of plant, displays and controls used by operators, admin. system
STEP 2 Review information from fault tree analysis	Check branches of fault trees for human failures affecting the top event
STEP 3 Talk-through	Familiarization with relevent procedures
STEP 4 Task analysis	Break down tasks into smaller discrete units of activity
STEP 5 Develop HRA event trees	Express each unit task sequentially as binary branches of an event tree. Each branch represents correct or incorrect performance (§3.2.2)
STEP 6 Assign human error probabilities	Data provided in the handbook (Swain and Guttmann, 1983)
STEP 7 Estimate the relative effects of performance shaping factors	Data provided in the handbook (Swain and Guttmann, 1983)
STEP 8 Assess dependence	Equations for modifying probabilities on the basis of dependence between tasks provided in the handbook (Swain and Guttmann, 1983)
STEP 9 Determine success and failure probabilities	Total probabilities for success and failures by multiplying branch probabilities and summing appropriately
STEP 10 Determine the effects of recovery factors	Operators may recover from errors before they have an effect. Recovery factors are applied to dominant error sequences
STEP 11 Perform a sensitivity analysis, if warranted	
STEP 12 Supply information to fault tree analysis	Human error probability or rate

For a plant at the design stage task descriptions must be developed from relevant detailed system information and preliminary procedures

Figure 27-1: Technique for Human Error Rate Prediction (THERP) method (CCPS, 2000)

Some human failures may depend on others and HRA should account for these dependencies. Plant personnel may also be able to recover from failures and recovery factors may be incorporated into the analysis. Uncertainty analysis is often performed to account for data and model uncertainties and sensitivity analysis is sometimes conducted to identify key factors and failures.

27.2 TOOLS

A variety of HRA tools exist, often known by acronyms such as THERP (Technique for Human Error Rate Prediction, HEART (Human Error Assessment and

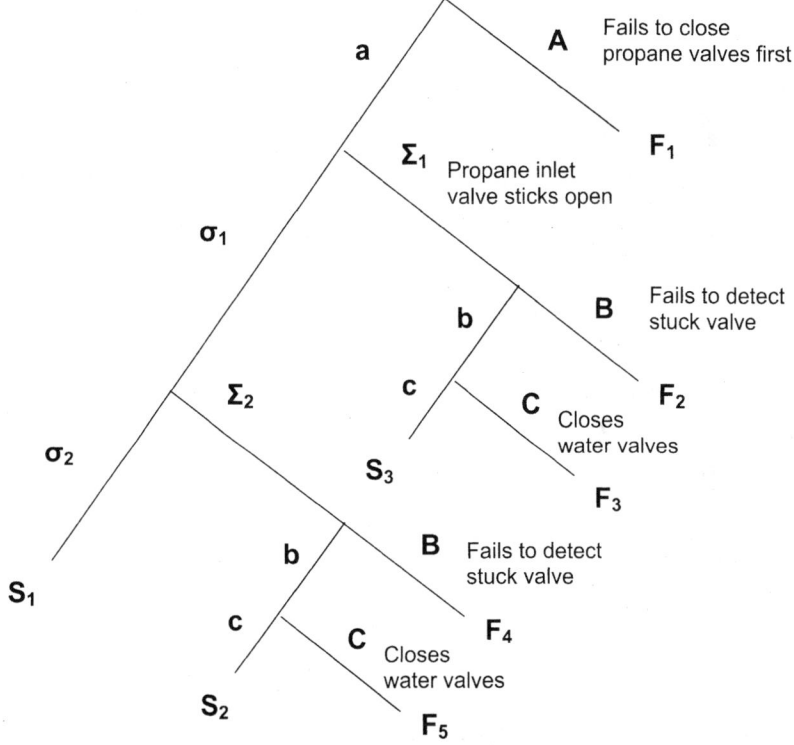

Figure 27-2: Example of an HRA Event Tree for a propane condenser. The potential human errors are represented by capital English letters and the potential equipment failures are represented by capital Greek letters. The series of events that will lead to the failure of interest is identified by a capital F at the end of the last branch in the event tree. All other outcomes are considered successes and are denoted by S. See "A Managers Guide to Reducing Human Errors; Improving Human Performance in the Process Industries" (Washington, DC: American Petroleum Institute, 2001). Note: This was originally published as "A Manager's Guide to Reducing Human Errors" (Washington, DC: Chemical Manufacturers Association, 1990).

Reduction Technique), SLIM (Success Likelihood Index Methodology), and IDEAS (Influence Diagram Evaluation and Assessment System), each with its pros and cons. The Technique for Human Error Rate Prediction (THERP) is perhaps the best known (Swain and Guttmann, 1983). THERP combines task and event tree analysis and was originally developed for the nuclear industry. A flowchart of the THERP method is shown in Figure 27-1. The development of HRA event trees is a key part of the method (see Figure 27-2). Other techniques either try to improve on THERP or simplify the analysis. Various texts describe the methods, their pros and cons, and their application in detail. They are listed in the reference section below. Draft guidelines have recently been published providing good practices for implementing HRA (NUREG-1792, 2004).

Data on human error rates are needed to perform the analyses. Published data are available (Swain and Guttmann, 1983). Error rates increase for various situations such as when people are called upon to perform complex tasks or respond to abnormal process operations when severe consequences are possible.

27.3 REFERENCES

CCPS (2000), "Guidelines for Chemical Process Quantitative Risk Analysis," 2nd ed. (NY: AICHE Center for Chemical Process Safety).

NUREG-1792 (2004), "Good Practices for Implementing Human Reliability Analysis (HRA)," NUREG-1792: Draft Report for Comment.

Swain, A. D. and Guttmann, H. E. (1983), "Handbook of Human Reliability Analysis with Emphasis on Nuclear Power Plant Applications" (Washington, DC: NUREG/CR-1278).

27.4 ADDITIONAL REFERENCES

Dhillon, B. S. (1988), "Human Reliability with Human Factors" (New York: Pergamon Press).

Dougherty, E. M. and Fragola, J. R. (1988), "Human Reliability Analysis" (New York: John Wiley).

Gertman, D. G. and Blackman, H. S. (1994), "Human Reliability and Safety Analysis Data Handbook" (New York: John Wiley).

Hollnagel, E. (1993), "Human Reliability Analysis, Context and Control" (New York: Academic Press).

Hollnagel, E. (1998), "Cognitive Reliability and Error Analysis Method (CREAM)" (Amsterdam: Elsevier Publishing).

Kirwan, B. (1994), "A Guide to Practical Human Reliability Assessment" (New York: Taylor and Francis).

Park, K. S. (1986), "Human Reliability: Analysis, Prediction, and Prevention of Human Errors," Advances in Human Factors/Ergonomics, Vol 7 (Amsterdam: Elsevier Publishing).

Salvendy, G. (Ed.) (1997), "Handbook of Human Factors," 2nd ed. (New York: Wiley-Interscience).

Sandom, C. and Harvey, R. S. (Eds.) (2004), "Human Factors for Engineers" (Savoy Place, UK: The Institution of Electrical Engineers).

CHAPTER 28

Safety Systems

28.1 INTRODUCTION

Various safety systems are used to control hazards in processes (CCPS, 1993a). These include instrumented protection, pressure relief and venting, release detection, isolation, secondary containment, off-gas treatment, fire and explosion protection, grounding, inerting, emergency shutdown, physical barriers, etc. Various human factors issues apply to the selection, design, construction, installation, testing, use and maintenance of these systems. Sometimes safety systems are taken for granted, and consequently they may not receive the attention required for their continued functionality.

28.2 PEOPLE AND SAFETY SYSTEMS

Safety systems may be inadequate, inoperative or disabled owing to various types of failures by people. A list of such human failures is provided in Table 28-1 and illustrated with examples. While none of these events will necessarily cause an accident immediately, they represent enabling events or conditions that may eventually allow an accident to occur.

Human factors that affect these failures must also be addressed. This includes lack of awareness, insufficient training, distraction, fatigue, carelessness, forgetfulness, inadequate procedures, mis-communication, work pressures, and conflicting objectives. These issues are similar to those for other aspects of processes but are especially important for safety systems. They apply throughout the life cycle of safety systems. This chapter focuses on human factors for some special aspects of safety systems. Human factors for other aspects in the life cycle of safety systems such as design, maintenance and testing, are similar to other equipment, albeit more important, and are described elsewhere in this book.

28.3 BYPASSING AND DISABLING SAFETY SYSTEMS

Operators or mechanics may deliberately disable or defeat an interlock system, trip, alarm or a shutdown system, or bypass a safety system, to perform maintenance,

Table 28-1: Human Failures for Safety Systems.

Failure	Example
Improper design	A storage tank dike is designed with inadequate height because the designer is not aware of the possibility of the liquid shooting out of an elevated hole and over the top of the dike.
Improper construction	The discharge of a relief valve is vented directly to the atmosphere instead of a scrubber owing to time pressures on the construction crew to finish the work under tight deadlines.
Improper installation	A check valve is installed backwards owing to carelessness by the mechanic.
Improper positioning	Gas detectors are placed at incorrect locations owing to miscommunication between engineering and maintenance.
Improper maintenance	A malfunctioning high temperature sensor is replaced but not connected by a distracted instrument engineer.
Improper testing	One of several oxygen analyzers is not tested by a tired mechanic and it subsequently fails to function correctly during a confined space entry.
Improper calibration	A pressure detector is calibrated incorrectly by a new employee who is not properly trained. It fails to register a pressure excursion.
Incorrect specification	A rupture disk is installed below a relief valve with an improper pressure rating owing to a slip by the mechanic.
Improper operation	A temporary replacement pump with a higher capacity than normal is connected to a tank fitted with a pressure/vacuum relief valve. Liquid is pumped out more quickly than air can enter through the vent and the tank collapses.
Deterioration from lack of regular maintenance	The filters on a scrubber are not cleaned or replaced because the maintenance department is overloaded and postpones the work in favor of activities deemed more important.
Deterioration from lack of PM	Grounding straps and cables become corroded because they are not included in the plant PM program due to an oversight.
Deterioration from lack of inspection	A dike wall loses its integrity over time since the slow deterioration is not noticed.
Failure to restore operation after maintenance	A fire protection system for a storage vessel is disabled to prevent its inadvertent operation while work is being performed. It is not returned to service after the work is complete due to lack of awareness, forgetfulness or inadequate procedures.
Deliberate disablement	A procedure calls for a blind to isolate a storage tank being taken out of service for maintenance. Operators decide to rely on remotely operated valves instead to perform the procedure in less time.
Failure to recognize the need for a safety measure	A tank truck containing a flammable liquid is offloaded using a non-conductive flexible hose. The driver was not aware of the static electricity accumulation hazard.

Note: Each type of failure may result from a variety of underlying human factors. The examples are intended to be illustrative only.

troubleshoot, tune a process, achieve higher production, alleviate an intermittent fault, or for convenience. Such disablements or bypasses may be intended as temporary but owing to forgetfulness, shift changes, or other reasons, restoration may not occur. In other cases they may be intended as a long-term solution to a problem such as nuisance alarms. Figure 28-1 shows the consequences of defeating a trip system that occurred in a hydrocracker in 1987.

Guidelines to address these human factors issues include:

- Ensure process designs do not encourage disablement. For example, alarm settings should not be close to set points for process parameters, otherwise they will be seen as nuisance alarms by operators and likely will be disabled.

- Do not allow bypassing of safety systems as a repeated solution to an underlying problem with the process. Fix the root cause of the problem.

- Establish a procedure for bypassing safety systems, including a review, and ensure it is followed.

- Set and enforce criteria for using bypasses.

- Require appropriate approvals for all bypasses, e.g. supervisor, engineering authority.

- Set and enforce time limits for bypasses.

- Require that operators keep and display a log of active bypasses.

Figure 28-1: The impact of disabling an unreliable low level trip on a high pressure separator—an explosion resulting in a fatality and $100 million of property damage.

- Use permits-to-work for bypasses. This helps ensure that system restoration occurs.
- Conduct periodic audits to confirm the correct treatment of bypasses.

28.4   SHUTDOWN SYSTEMS

Many processes use safety shutdown systems as a last resort in the event that other control and safety measures do not work. They are often automated but sometimes they are manual. Important human factors issues can arise when operators are called upon to make a decision on shutting down a process. Process dynamics and time factors may require decisions to be made by operators without consulting supervisors or managers. Shutdown procedures must address the level at which a shutdown decision can be made.

Shutdown decisions can be influenced by a variety of factors besides the risk of not shutting down. For example, operators may be disinclined to shut down in order to avoid production losses, personal inconvenience, loss of their reputation or self respect, or negative feedback from peers and managers. Peer pressure, concerns about job security, fear of recrimination, lack of understanding of the consequences, belief that normal operations can be restored without shutting down, or the desire to be seen as a hero may also be factors.

Guidelines to address these issues include:

- Use automated shutdown systems where possible to take the decision out of the operators' hands.
- Establish as clearly as possible criteria for when processes must be shut down manually. Provide clear management direction to shut down processes in these circumstances.
- Ensure that operators understand abnormal process operations and the consequences of deviations from normal operation.
- Train operators in emergency shutdown and provide refresher training so that operators feel comfortable and familiar with the process when emergency shutdown must actually be performed. ·
- Ensure that management reaction in the event of emergency shutdown by operators provides positive reinforcement of the operators' actions.
- Establish a strong process safety culture. Ensure that when in doubt, a shutdown decision is made.

A process plant producing a highly exothermic polymer product contained a number of emergency shutdown stations scattered throughout the plant. All operating personnel were given the authority to use the shutdown stations, even if they had only a hunch that something was wrong.

An inquiry was made of the plant manager as to how frequently this occurred. The plant manager replied: "I don't know and it's none of my business!"

28.5   SELECTION OF SAFETY SYSTEMS

Various human factors enter into the selection of safety systems. Engineers must fully understand process hazards in order to select appropriate safety systems. For example, antistatic plastic or paper bags may be needed when handling flammable dusts. Hazard identification and analysis play a key role in helping ensure process hazards are understood. Engineers must be informed of the results of these studies.

Engineers must also select the right type of safety system. For example, insulated dike walls may reduce the volatilization of liquids spilled at temperatures above their boiling point. There can also be a tendency for engineers to add safety systems beyond the point of diminishing returns. In particular, this can result from performing process hazards analysis and indiscriminately adding safeguards. The more safety systems that are installed, the more complexity there is in the process, the greater the possibility that new hazards have been introduced and the more can go wrong. Engineers should not decide on the installation of safety systems in isolation. They should consider possible interactions between safety systems that may increase risk or introduce new hazards. Engineers should also consider the impact on process safety of the susceptibility of safety systems to human failures, e.g. manual vs automatic isolation valves.

Pointers to address these issues include:

- Ensure engineers are informed of process hazards.
- Decisions on safety systems should be based on PHA and risk analysis.
- LOPA should be used to provide objectivity for decisions on the need to add safety systems.

28.6   COMMON CAUSE FAILURES IN SAFETY SYSTEMS

Common cause failures arise when a single failure results in multiple simultaneous failures in systems or components that would otherwise be expected to fail independently (CCPS, 2000). Safety systems often are designed to provide redundancy with duplicate safety systems provided. However, the redundancy will be lost if common cause failures occur. For example, if a mechanic is tasked with calibrating three identical pressure detectors but systematically mis-calibrates all three detectors in the same way owing to lack of familiarity with the calibration procedure, then all three detectors will fail to correctly detect pressure excursions. The detectors all fail for a common cause. Such failures are important because they result in the likelihood of simultaneous multiple failures being significantly higher than what would be expected if the failures occurred independently of one another.

People are a common cause of common-cause failures and they are involved in all steps of the life cycle for safety systems—design, specification, construction, installation, maintenance, testing, etc. Common cause failures can occur in all these steps. One defense against common cause failures is to provide diversity for the safety systems protecting a process so the systems are less susceptible to failure from the same cause. The same principle can be applied to the people involved with a process to lessen the chance for common-cause failures that may originate with one individual. This can be accomplished by not assigning the same individual to carry out activities susceptible to common cause failure or by providing one-over-one checks for critical actions in design, construction, maintenance, etc. of safety systems.

Common cause failures should be considered in process hazard analysis. More detailed studies can also be performed, e.g. using fault tree analysis (CCPS, 2000).

A procurement specialist in a company handling anhydrous ammonia was charged with ordering some process components. He noticed in his ordering system that cheaper components were available but did not understand that they were for aqueous ammonia service and not anhydrous service. The components were ordered and installed and multiple leaks of anhydrous ammonia occurred.

28.7   TOOLS

The starting point to address human factors for safety systems is an understanding of the types of human failures that are possible. Table 28-1 provides a checklist of the types of human failures that should be considered for safety systems. Human factors for the design, construction, installation, maintenance, and operation of safety systems are similar to those for other process equipment. These human factors issues should be addressed in process hazard analysis (see Chapter 27) and process safety audits (CCPS, 1993b) with emphasis on potential common cause failures.

Various other chapters provide tools that should also be applied to safety systems. This includes Process Equipment Design (Chapter 4), Environmental Factors (Chapter 15), Procedures (Chapter 22), Training (Chapter 12), Competence Management (Chapter 29), Management of Change (Chapter 25), Maintenance (Chapter 23), and Communications (Chapter 13). Issues such as work pressures and conflicting objectives should be addressed by the organization's management system. Guidelines for addressing safety-system-specific issues have been provided in earlier sections of this chapter.

28.8   REFERENCES

CCPS (1993a), "Guidelines for Engineering Design for Process Safety" (NY: AICHE Center for Chemical Process Safety).

CCPS (1993b), "Guidelines for Auditing Process Safety Management Systems" (NY: AICHE Center for Chemical Process Safety).

CCPS (2000), "Guidelines for Chemical Process Quantitative Risk Analysis" (NY: AICHE Center for Chemical Process Safety).

CHAPTER 29

Competence Management

29.1 INTRODUCTION

The *competence* of personnel is defined as the ability of personnel to perform tasks according to expectations. Competence is crucial to every organization because of the role it plays in ensuring tasks are carried out satisfactorily and safely. Competence implies appropriate qualifications, training, skills, physical and mental capabilities, knowledge, understanding, behavior and attitudes as well as the ability to perform tasks according to defined performance standards, as shown in Figure 29-1. Often, assumptions of competence are made based on the adequacy of experience or training, possession of qualifications, or the availability of a procedure. Not only may such assumptions be incorrect, but they do not capture all the key aspects of competence.

Accidents have occurred when competencies other than knowledge and experience were absent (HSE, 2003). Competence management is particularly important when an organization relies heavily on the skills, knowledge and capabilities of its personnel, such as in the process industries. The importance of competence is compounded in the face of the re-engineering, downsizing, and multi-tasking that can occur in the process industries. Personnel are often expected to take on a wider range of tasks with less supervision, thus increasing the need to manage competence effectively. Competence management should be planned, proactive and systematic. It involves:

- Identification of competence requirements
- Selection and recruitment of personnel
- Assessment of competence
- Certification of competence
- Maintaining, reassessing and monitoring competence

Benefits of competence management include:

- Identifying gaps in personnel competencies before they contribute to accidents.
- Reducing the likelihood of overlooking substandard personnel competencies.
- Determining training needs.

- Validating and improving training.
- Justifying reliance on personnel rather than supervision, written instructions or procedures.
- Assuring stakeholders that a key aspect of safety performance is being managed.
- Contributing to continuous performance improvement.

Competence management should be part of an organization's overall management system. It should apply to personnel from the top to the bottom of the organization and include contractors and others who perform work within the organization.

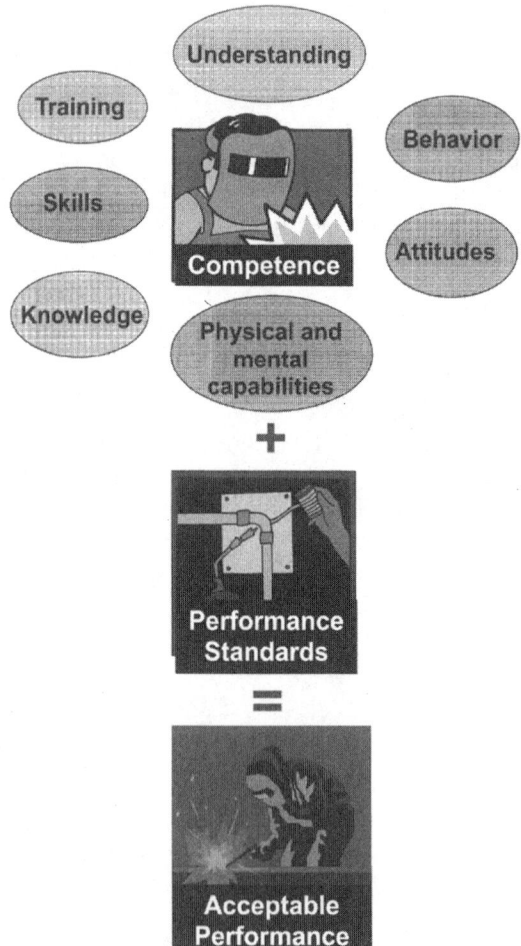

Figure 29-1: Competence management. Copyright © 2005, Primatech Inc. All rights reserved.

OSHA uses the term "competent person" in many standards and documents. On the OSHA web site (OSHA, 2005) a competent person is described as an individual who, by way of training and/or experience, is knowledgeable of applicable standards, is capable of identifying workplace hazards relating to the specific operation, is designated by the employer, and has authority to take appropriate actions. Some OSHA standards define additional requirements that must be met by the competent person.

29.2 ELEMENTS OF COMPETENCE MANAGEMENT

Identification of Competence Requirements

The focus should be on safety-critical tasks/jobs. All aspects of the process life cycle including operations and maintenance should be considered. Competencies required for personnel to perform tasks/jobs should be defined, together with assessment criteria, as part of job descriptions. This includes qualifications, training, skills, physical and mental capabilities, knowledge, understanding, behaviors, attitudes and performance standards. They should be realistic and appropriate for the tasks/jobs performed. Competence requirements for a task/job may be graduated for varying levels of competence such as supervised practitioners, practitioners and experts.

Various tools are available for use in identifying competence requirements including:

- Task analysis
- Task descriptions
- Skill and knowledge inventories
- Training needs analysis
- Job hazard analysis
- Process hazard analysis
- Risk assessment

Selection and Recruitment of Personnel

Organizations should ensure that individuals are selected and recruited who possess or are capable of acquiring the appropriate competencies to perform assigned tasks/jobs. Competence-job matrices are sometimes used to define requirements. Selection of personnel is based on whether an individual has the aptitude and appropriate characteristics for a job. Aptitude tests and psychometric personality tests are sometimes used. The ability to perform the job competently is usually developed through training and experience. This addresses gaps in personnel competencies compared to requirements. Personal development plans are prepared to address these gaps and may include, for example, self study, formal training, professional development, or on-the-job training.

Initial Training and Development

Competence to perform a job is often developed through initial training followed by coaching and supervision by experienced personnel. Training is discussed in Chapter 12.

Assessment of Competence

Competence should be assessed before personnel are allowed to perform unsupervised work. Methods used should be appropriate to the task and the risks involved and must be valid and reliable. Verbal or written tests, observation of job/task performance, and behavioral observation are used. Simulators are used in some industries, particularly for infrequent events including emergency situations, process upsets, and operations such as start-up and shutdown. The assessment of competence should be correlated with subsequent job performance to validate the method used.

Assessors should themselves be competent in assessing competence as well as be credible, consistent and independent.

Certification of Competence

Organizations should formally designate personnel as meeting required competencies. This may be in the form of a simple certificate or other documentation. In some cases, personnel may be certified, licensed or accredited by external organizations. Care should be exercised to ensure such certification, licensing or accreditation is meaningful for the actual tasks and jobs to be performed.

Maintaining, Reassessing and Monitoring Competence

Personnel development should proceed according to a plan prepared for each individual. This should include refresher training provided at a frequency based on the anticipated deterioration in competence. Particular attention should be paid to maintaining skills needed to handle infrequent events. The provision of procedures is also an important part of helping to ensure tasks are performed consistently and correctly (see Chapter 22). Other forms of assistance such as supervision, coaching and job aids should be provided, as appropriate.

Competencies should be re-assessed periodically such as through performance appraisal. Actual performance may be reviewed or other forms of assessments, such as demonstrations, may be used for tasks not performed during the appraisal period. The frequency of reassessment should be based on the safety criticality of the task, its frequency of performance, and the anticipated decay rate for competencies. There must be suitable responses in the event of sub-standard performance, for example, improvement in training, personnel selection, etc. Reassessment should also be performed to qualify an individual to return to the performance of critical tasks after an absence.

Table 29-1: Engineering skills targets by discipline

	Skills Targets by Disciplines					
	Project Managers	Project Engineers	Process Engineers	Cost/ Scheduling	Control Systems	Engineering Technologists
Stage 1: Project Initiation						
Inherently Safer Technology	W	W	P	A	W	P
Stage 2: Technology Selection and Development						
Material Technical Sheets	A	A	P	A	A	P
Reaction Sheets	A	A	P	A	A	P
Interaction Matrices	A	A	P	A	A	P
Preliminary Safety Analysis	W	W	P	A	A	W
Siting (Layout)	P	P	W	A	W	A
Human Factors	P	P	W	A	P	A
Stage 3: Project Definition						
Material & Energy Balance	W	W	P	A	W	W
Basic Engineering Safety Review	W	W	P	A	W	W
Stage 4: Definition of Facilities						
Layers of Protection Analysis	W	W	P	A	W	A
Risk Sheets	W	W	P	A	W	A
Stage 5: Engineering and Procurement Construction Activity						
Management of Change/Design Changes	P	P	P	A	W	A
What-if Review	P	P	P	A	W	A
Pre-Startup Safety Review	W	W	W	A	W	A
Mechanical Integrity	P	P	W	A	P	A
Hazard Identification						
What-if	P	P	P	A	W	A
HAZOP	P	P	P	A	W	A
FMEA	A	A	A*	A	A	A
Fault Tree	A	A	A	A	A	A
Technical						
Safety Instrumentation (SIS/SIL)	W	W	P	A	E	A
Pressure Relief Sizing	A	A	P	A	A	E
Two-phase flow analysis	A	A	W/P	A	W	E
Consequence Analysis/Dispersion Modeling	A	A	W	A	A	P/E
Dust and Gas Explosion Analysis	A	A	W/P	A	A	P/E
Static Electricity	A	W	W	A	A	P/E
Adiabatic Calorimetry/Runaway Reactions	A	A	W	A	A	P/E
Fire Detection, Prevention and Protection	A	W	A	A	A	A
Release Mitigation (flares, scrubbers, etc.)	A	W	P	A	A	P/E
Regulatory/Policy						
Process Hazard Management Program	A	A	A	A	A	A
Company Design Standards	P	P	P	P	P	W
Health & Safety Basics	W	W	W	A	A	A

(continued)

Table 29-1: Engineering skills targets by discipline *(continued)*

	Skills Targets by Disciplines					
	Project Managers	Project Engineers	Process Engineers	Cost/ Scheduling	Control Systems	Engineering Technologists
Regulatory/Policy *(cont.)*						
Environmental Basics	W	W	W	A	A	A
Audit Programs	A	A	A	A	A	A

A = Awareness, W = Working knowledge, P = Proficiency, E = Expert.

Awareness: Individual has basic knowledge in the particular functional area. Individual should work under the direction and guidance of more experienced practitioners.

Working Knowledge: Individual has developed expertise and knowledge to a level that allows them to work independently within a given skill area and handle routine issues. Oversight and mentoring by an expert is warranted for complex or new projects within a given skill area. Working knowledge generally involves background training and education plus experience and mentoring for each skill area.

Proficiency: Individual is acknowledged by the functional expert as achieving a high degree of knowledge and expertise for application in a given area. The individual can respond independently on most issues. Generally they would seek input from an expert for specialized issues where a new or complex situation is encountered.

Expert: Individual is acknowledged by Process Safety functional members, Site and Business leaders as having a high level of proficiency in given skill areas. They would provide coaching, guidance, direction and development activities for the functional skill area.

Trends and developments in competency levels over time should be tracked in order to determine if systemic problems develop in the competence management program and to permit continuous improvement.

A refining company used a contractor for maintenance. The contractor was charged with performing a cleaning operation using a high-pressure water hose. An unqualified employee of the contractor was assigned the task. He was unable to properly control the hose and was severely injured.

29.3 TOOLS

The UK Health and Safety Executive has published a report (HSE, 2003) that provides a review of current competence assessment practices, an approach for competence assessment in relation to major accident prevention, and advice, self-assessment checklists and examples of competence assessment as benchmarks. The report has drawn together experience, standards and lessons learned from a number of high hazard industries, particularly chemicals, onshore and offshore oil, and nuclear and aviation. It also considers guidance on competence assessment from personnel specialists, certification bodies and institutes.

The UK Institute of Electrical Engineers has published voluntary guidelines

(IEE, 1999) that cover competency requirements for practitioners working with electrical, electronic and programmable electronic (E/E/PE) systems. These guidelines follow from the broadly defined competencies on functional safety in IEC 61508 by setting out a procedure for assessing competencies of personnel working on safety-related tasks involving E/E/PE systems.

Various consulting companies offer commercial tools, including software products, that can be used in competence management. Current tools can be identified through a web search.

Table 29-1 provides an example of a skills matrix based on discipline used by one company.

29.4 REFERENCES

HSE (2003), Research Report 086, "Competence Assessment for the Hazardous Industries" (Sudbury, UK: HSE Books).

IEE (1999), "Safety, Competency and Commitment: Competency Guidelines for Safety-Related System Practitioners" (London, Institute of Electrical Engineers).

OSHA (2005), Web site: http://www.osha.gov/SLTC/competentperson/index.html

29.5 ADDITIONAL REFERENCES

Fletcher, S. (2001), "Competence-Based Assessment Techniques" (London: Kogan Page Publishers).

Rothwell, W. J., Hohne, C. K., and King, S. B. (2000), "Human Performance Improvement, Building Practitioner Competence" (Oxford, UK: Butterworth-Heinemann, now Elsevier).

Spencer, S. M. and Spencer, L. M (1993), "Competence at Work: Models for Superior Performance" (Hoboken, NJ: Wiley).

Zwell, M. (2000), "Creating a Culture of Competence" (Hoboken, NJ: Wiley).

CHAPTER 30

Emergency Preparedness and Response

30.1 INTRODUCTION

It is vitally important to prepare for emergency or abnormal situations prior to their occurrence. Humans play a dominant role during emergencies and human behavior during stressful situations is often unpredictable. For example, the US Nuclear Regulatory Commission (NRC, 1975) once estimated that even highly-trained nuclear reactor operators would do the wrong thing more than 9 times out of 10 in the first minute after a nuclear reactor emergency shutdown.

Human factors plays a very important role in emergency preparedness and response. Preparing for the emergency or abnormal situation prior to its actual occurrence will increase the likelihood that the human actions will mitigate the abnormal situation. The preparation should identify likely emergency scenarios, develop a simple plan to handle the emergency, and then practice those simple plans.

30.2 TOOLS

Study Scenarios and Training

The first step to prepare for an emergency or abnormal situation is to study and create credible, written scenarios. The scenarios are used to develop action plans and emergency procedures, and to ascertain and mark egress routes, plan the training, and create simulations. It is extremely important that these scenarios and subsequent documents and training be updated and reviewed on a regular basis to ensure that system changes are adequately considered.

Understand the System

Humans must process a vast amount of information and arrive at the correct decision. To do this reliably, a person must understand the system and its status during the event. Training, as discussed previously, is used to ensure that the person understands the system. The current status of the system requires several things—

this includes the use of a large situation awareness display in the control room, local and centralized alarms with adequate description, and appropriate event procedures. Because emergency response events will always occur at the worst possible time, it is important to have a visible structured process for plant operators to follow. An example of a logic diagram to aid the plant in insuring that necessary communications are completed is shown as Figure 30-1. By following a defined logic process there will be less chance for human error in a high stress emergency event.

Incident Command Systems

Incident command teams are a common ad-hoc team specifically organized to deal with an abnormal event. These are used by fire and police departments and many Fortune 500 companies. They involve a pseudo-militaristic structure to establish a command hierarchy and provide intra-team communications. This structure allows fire-fighters and other first responders to use their skill-based techniques to deal with situations that are immediately in front of them and which could change second-by-second. It also allows others (higher) in the command structure to use rule and knowledge-based practices to focus on diagnosing patterns, taking preventive or precautionary measures, etc. The needs of the two are quite different: firefighters need things immediately at hand, clearly labeled and often painted in bright colors; incident command needs excellent communications, a steady stream of information to make good decisions, and some degree of calmness to allow clear thinking. A company or site that practices with an incident command structure similar to their local emergency responders will integrate better in case of a large emergency.

Drills

Drills should be carried out with sufficient frequency so that operators can understand emergency situations and react properly. These exercises must be realistic, using the same equipment, communications, and time pressures. These should be used to train people and measure their ability to react appropriately.

Emergency Response Equipment (Selection and Maintenance)

Equipment necessary for abnormal events must be on hand and ready for use. Such equipment may include respirators, HAZMAT suits, fire-resistant suits, extinguishers, sprinkler systems, trucks, radios, and many others. Some of this equipment has regulatory requirements, including respirators, fire extinguishers, and sprinkler systems.

Stores of emergency response equipment must be inventoried and inspected routinely to ensure that it exists and is in good condition. Some equipment may have a shelf life and may need to be replaced if outdated.

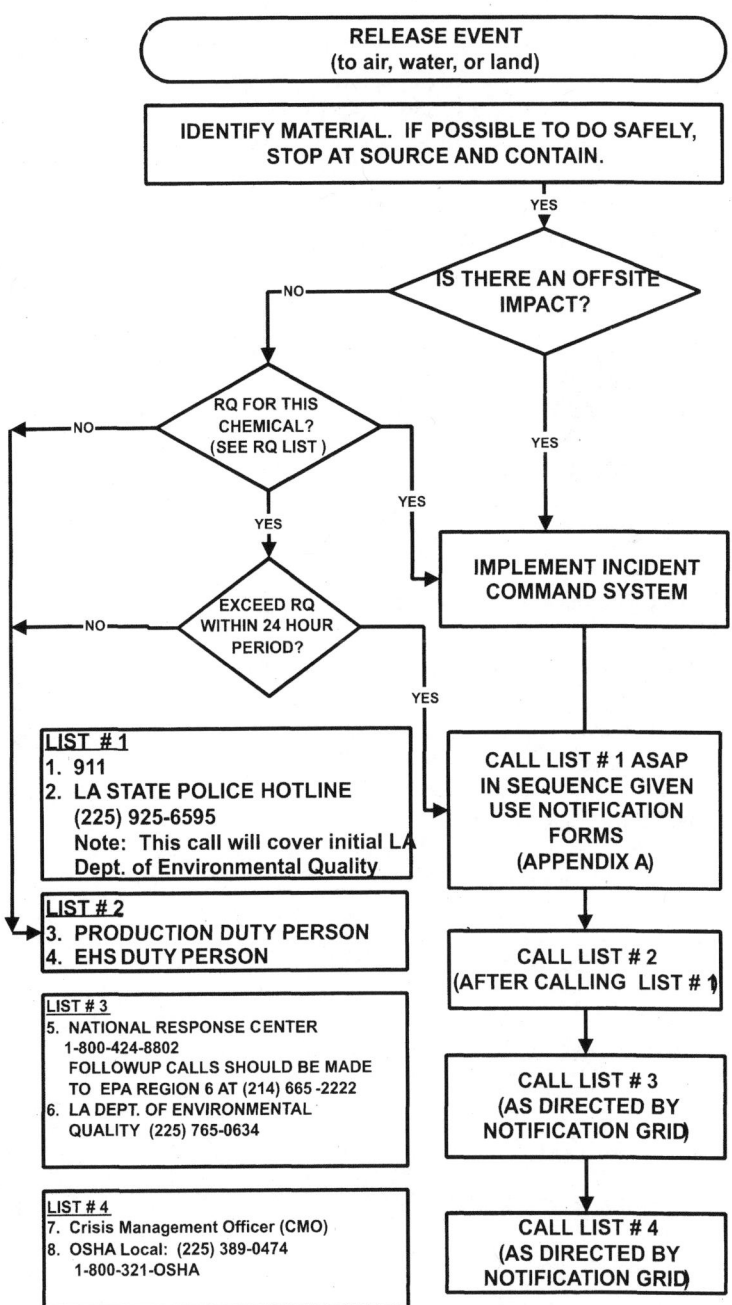

Figure 30-1: A call-down flowchart for a chemical release event.

A company added foam capability to its fire monitors in an area where flammable hexane was stored in diked vertical storage tanks. They trained their fire crews to use the foam in case of large leaks to suppress vapors being released from the liquid. About a year later, a pressure storage vessel holding butane was overpressured and its relief valve lifted to atmosphere. The fire crew used their previous training and used foam which was completely inappropriate. Fortunately, no one was hurt and no damage occurred.

The company explained to the fire-crew leaders what the purpose of the foam was, and developed a simpler rule for when to use foam.

30.3 REFERENCE

NRC (1975), "Reactor Safety Study: An Assessment of Accident Risks in US Commercial Nuclear Power Plants" (Washington, DC: U.S. Nuclear Regulatory Commission).

30.4 ADDITIONAL REFERENCES

Erickson, P. (1999), "Emergency Response Planning for Corporate and Municipal Managers" (Amsterdam: Academic Press, now Elsevier).

HSE (2001), "Performance Indicators for the Assessment of Emergency Preparedness in Major Chemical Accidents (London, UK: U.K. Health and Safety Executive).

OSHA (2002)), "1910.38: Emergency Action Plans" (Washington, DC: U.S. Occupational Safety and Health Administration).

OSHA (2004), "Principal Emergency Response and Preparedness Requirements and Guidance" (Washington, DC: U.S. Occupational Safety and Health Administration).

CHAPTER 31

Incident Investigation

31.1 INTRODUCTION

Virtually all incidents can be attributed to human error—either directly causing the incident or failing to properly respond to it. Thus, it is essential that any incident investigation program consider human factors.

31.2 ISSUES/EXAMPLES

Unfortunately, after an incident, the tendency is to blame the individual for the error, and the obvious solution is to retrain, reassign, counsel, discipline, or fire that person. However, as shown in other sections of this book, human error is seldom simply a matter of individual misconduct. In most cases, there are reasons for the behavior that led to the incident, and they are rooted in the design of the equipment, the design of the task, or the design of the work environment (including both the physical and organizational aspects).

The challenge in incident investigation is to identify those underlying factors that led to the observed error(s). This is an organizational culture issue. Some organizations conduct their incident investigations like criminal investigations—find the perpetrators and punish them. Management believes that this will change the behavior of the individual involved and serve as stark warning to others to be more attentive or they will suffer the same fate. As an added bonus, by pinning the blame on the individual worker, management can absolve itself of any blame in the matter and congratulate itself on strong action that will prevent future incidents. Sadly, this approach is doomed to failure, and future incidents are inevitable.

A more successful approach is to realize that virtually all incidents are symptoms of a breakdown in the management systems. The emphasis is on learning from near misses and correcting the problems before an incident occurs (CCPS, 2003; HSE, 2004; NRC, 1992). However, for this approach to be successful, workers must trust that management will not use self-reported errors as an excuse to punish individuals today, or at some unknown point in the future. Thus, for example, in the aviation community, pilots self-report errors to the National Aeronautics and Space Administration (NASA), which extracts the lessons learned while protecting the individual pilots' identities (FAA, 1997). NASA then reports the lessons learned to the Federal Aviation Administration, which acts to improve the air traffic system. Companies wishing to adopt this model, especially those with a history of the crime-and-punishment approach, may need to engage a third party as the intermediary to gather information confidentially.

Assurances of confidentiality alone are necessary, but not sufficient, to change an organization's culture. It is always easier to ignore a near miss than to report it. Therefore, you should make reporting a near miss easy for each individual and should consider offering incentives to encourage participation in the system. These can be as simple as a monthly drawing for movie tickets as a small token of the company's appreciation or a more substantial reward periodically for those reports that had the greatest learning value to the organization. But the most powerful incentive of all is for the company to respond to the reports and make substantial, visible improvements. Workers then see that their input is valuable and that management is actively working to help them do their job safely and productively.

To have an effective program, an organization must also be sensitive to human factors issues that will arise during the investigation, especially during data collection. As soon as possible, anyone with information about events before, during, or after an incident should write down their recollections. This minimizes data loss from forgetfulness and from contamination by the opinions of others. Soon thereafter, one or two members of the incident investigation team should interview each witness. The objective is to gather all relevant facts, and witnesses are generally more cooperative if they do not feel threatened. The interviewer can maximize his effectiveness by:

- Being a peer of the interviewee
- Being uninvolved in the incident
- Conducting the interview at the scene of the incident (to aid recollection and demonstration) or at a neutral location
- Providing photographs, drawings, procedures, etc. to aid the witness
- Introducing himself and the meeting objectives
- Establishing and maintaining rapport with the witness
- Recording meeting notes on paper instead of audiotape or videotape
- Asking open-ended questions
- Avoiding arguments and criticism
- Reviewing and confirming the accuracy of meeting notes with the witness
- Getting permission for follow-up interviews, if necessary

If the interviews reveal that human errors caused or contributed to the incident, the investigation team can use the topics in this book as a checklist to aid in their search for the underlying deficiencies in the work situation that led to the error(s). By correcting these deficiencies, the plant can reduce the likelihood of recurrence far more effectively than by simply punishing the individual(s) involved.

A maintenance worker was asked to change a gasket on a pump. The work permit clearly stated that gloves and goggles were to be worn. Unfortunately, the worker did not wear the required goggles and chemical splashed into his eyes. Further investigation showed that all of the work permits at this facility stated that gloves and goggles were to be worn, even for tasks that did not require them. This sent a conflicting message to the employees: If management does not wish to make the effort to establish the need, then perhaps it is not important.

31.3 TOOLS

Auditing is the primary tool for insuring that human factors are considered in incident investigations. Are the corrective actions focused on the individuals involved—retraining, coaching, time off—or are they focused on correcting the root causes of the human errors? What is the ratio of near-miss investigations to incident investigations? A healthy investigation program should have a ratio of at least 10:1, and any ratio less than 1:1 indicates an investigation program that is missing enormous opportunities to improve human factors.

The Management Oversight & Risk Tree (MORT) tool (Johnson, 1973) was developed by the U. S. Department of Energy for the investigation of incidents. It is a predefined tree structure that guides the user to identify weaknesses in safety management systems, many of which directly identify human factors issues that should be improved. There are also many proprietary tools for structured root cause analysis that have been developed from MORT.

31.4 REFERENCES

CCPS (2003), "Guidelines for Investigating Chemical Process Incidents," 2nd ed. (New York: AICHE Center for Chemical Process Safety).

HSE (2004), "Investigating Accidents and Incidents—A Workbook for Employers, Unions, Safety Representatives and Safety Professionals" (HSG245) (Sudbury, UK: Health and Safety Executive).

NRC (1992), "Incident Investigation Manual," Rev. 2 (NUREG-1303) (Washington, DC: U.S. Nuclear Regulatory Commission).

FAA (1997), "Aviation Safety Reporting System" Advisory Circular 00-46D (Washington, DC: Federal Aviation Administration)

Johnson, W. G. (1973), *MORT—the Management Oversight & Risk Tree* (SAN 821-2), (Washington, DC: U.S. Atomic Energy Commission).

Human Factors Checklist

The following checklist distills many of the key points in this book and is organized to parallel the human factors tool kit. Refer to Section III for further discussion of the issues the questions are designed to reveal. Ideally, designers contemplating new or modified facilities, equipment, or procedures will use the checklist, as will those attempting to identify human factors issues in existing facilities.

Those questions with an "S" following the item number can be evaluated in a static condition—by looking at drawings, photographs, or physical equipment. Those questions with a "D" following the item number should also be evaluated under dynamic conditions—during actual or simulated use by workers.

Analysis of _____

at _____ Facility

Date Completed:_____   Completed By:_____

Item S/D*	Question	Response (Y/N/NA)	Comments
	4. Process Equipment Design		
4.1 S	Is equipment suitable for the task?		
4.2 S	Does the equipment design avoid unnecessary complexity?		
4.3 S	Is the equipment easily found and recognized?		
4.4 S	Is the equipment easily accessible?		
4.5 S	Is the equipment easily seen or heard?		
4.6 S	Is the equipment available?		
4.7 S	Are the organization, arrangement, and operation of equipment logical and consistent?		
4.8 D	Is the equipment adjustable to the user?		
4.9 S	Does the equipment conform to user expectations?		
4.10 S	Has the human-process interface ever undergone a human factors analysis?		
4.11 S	Is there a formal mechanism for correcting human factors deficiencies identified by the operators?		
4.12 S	Are designers made aware of human factors problems so they can improve future designs?		
4.13 D	Are normal and emergency operations within the physical capabilities of operators (none require excessive force by the operator)?		
4.14 D	Has the operator's mobility been considered in determining what protective gear is required for certain tasks, including emergency response?		
	5. Process Control Systems		
5.1 S	Is the control system reliable?		
5.2 S	Can operators force control inputs into a desired state?		
5.3 D	Does the computer check that values entered by operators are within a valid range?		
5.4 D	Can operators safely intervene in computer-controlled processes?		
5.5 D	Are automatic response features incorporated when a process upset/ condition requires rapid response?		

*S = static conditions; D = dynamic conditions.

Human Factors Checklist

Analysis of _____

at _____ Facility

Date Completed:_____ Completed By:_____

5.6 D	Are automatic features provided when a process upset/condition may be difficult to diagnose in a timely manner due to complicated processing of information (requiring a knowledge-based decision)?		
5.7 D	Are there initiation cues for the task or step: a. Does equipment design support the initiation of the task? b. Is there a procedure that initiates the task? c. Have the operators been trained on the meaning of the initiating cue?		
5.9 D	Are the cues unambiguous and distinctive from the other signals? a. Are the controls and displays consistent with human factors design standards? b. Are the controls and displays consistent with the populational stereotypes of the group using the system?		
5.10 D	Is the desired action clear?		
5.11 D	Can the operator determine the current status of the system versus the desired state?		
5.12 D	Can the operator correct the situation or perform the action within the required time period?		
5.13 D	Is the feedback on control action prompt and direct?		

6. Control Center Design

6.1 S	Does the layout provide adequate access, egress, and freedom of movement?		
6.2 S	Does the layout provide adequate lines of sight?		
6.3 S	Does the layout facilitate necessary communications?		
6.4 S	Does the layout provide sufficient space for workers and their workstations?		

7. Remote Operations

7.1 S	Is there adequate communication with field operators?		
7.2 D	Can remote operators adequately coordinate with field operators?		
7.3 D	Do workers in remote locations have adequate displays of operating information?		
7.4 D	Do workers in remote control rooms see and hear the equipment they control (camera, microphone)?		
7.5 D	Do other activities distract workers at remote locations?		
7.7 D	Is the travel time from the control room to the unit acceptable?		
7.8	Do operators in remote locations periodically		

*S = static conditions; D = dynamic conditions.

Human Factors Checklist

Analysis of _____

at _____ Facility

Date Completed: _____ Completed By: _____

S	spend time in the field?		

8. Facilities and Workstation Design

8.1 S	Do workplaces accommodate the extremes of the user population?		
8.2 S	Do workplaces adjust to the characteristics of the user population?		
8.3 S	Is there adequate access for routine operation and maintenance of all equipment?		
8.4 D	Does the equipment avoid forcing human joints beyond the range of natural movement?		
8.5 D	Can workers avoid holding tensed muscles in fixed positions for long periods?		
8.6 D	Are frequently accessed items within easy reach?		
8.7 S	Are workers' hands located near elbow height?		
8.8 S	Are repetitive tasks minimized?		
8.9 D	Do frequent tasks avoid high contact forces?		
8.10 S	Are specialized tools provided to reduce body stress?		
8.11 S	Are workstations and seating arranged according to human factors standards?		
8.12 D	Can workers move easily between sitting and standing positions?		
8.13 S	Is equipment designed for ease of maintenance?		
8.14 S	Are the controls and displays arranged logically to match the expectations of the operators?		
8.15 S	Are the controls and displays adequately visible from all relevant working positions?		
8.16 S	Is all significant operating information logically arranged?		
8.17 S	Are related displays and controls grouped together?		
8.18 S	Are the controls distinguishable and easy to use?		
8.19 S	Do the displays provide adequate information to assess the status of the entire process as well as essential details of individual systems?		
8.20 S	Do the displays and indications support the procedural requirements?		
8.21 S	Have the operators made any modifications to the displays, controls, or equipment to better suit their needs?		
8.22 S	Do controls conform to strong populational stereotypes (color, direction of movement, etc.)?		
8.23 S	Do the control panel layouts reflect the functional aspects of the process or		

*S = static conditions; D = dynamic conditions.

Human Factors Checklist

Analysis of _____

at _____ Facility

Date Completed:_____ Completed By:_____

	equipment?		
8.24 D	Does the control arrangement logically follow the normal sequence of operation?		
8.25 S	Is there a dedicated Emergency Shutdown panel and is it located on an egress route?		
8.26 D	Can the operators respond to upset/emergency conditions in a timely manner?		
8.27 D	Can operator and maintenance workers safely perform all required routine and emergency actions given the physical arrangement of the equipment (e.g., proximity of tasks to hazards, such as rotating equipment, hot surfaces, or vented equipment)?		
8.28 D	Is equipment that requires urgent manual adjustment or actuations (e.g., emergency shutdowns) easily identifiable and readily accessible?		

9. Human/Computer Interface

9.1 S	Do screens use familiar metaphors (process structure, process functions, tasks, etc.)?		
9.2 S	Is text legible (font size and type)?		
9.3 S	Is information presented suitably (numeric value, mimics, tables, trends, bar graphs, etc.)		
9.4 S	Are symbols, text, and numerical information presented appropriately?		
9.5 S	Does the screen layout appropriately use white space, emphasis, and fonts?		
9.6 S	Is there adequate spacing between active areas of touch screens?		
9.7 S	Are screens and information readily identifiable (use of titles, labels, and icons)?		
9.8 S	Is information displayed in a format adapted to the operator?		
9.9 S	Is appropriate and relevant information presented?		
9.10 S, D	Do screens provide only the information the operator needs (not excessive detail)?		
9.11 S	Is information located in an appropriate and logical manner?		
9.12 S, D	Is the most important and frequently used information most prominently displayed?		
9.13 S	Do screen schematics correlate with the actual plant configuration?		
9.14 S	Are there text labels for icons?		
9.15 S	Is appropriate coding used (shape, color, alphanumerics)?		

*S = static conditions; D = dynamic conditions.

Human Factors Checklist

Analysis of _____

at _____ Facility

Date Completed:_____   Completed By:_____

9.16 S	Do color choices match cultural expectations?		
9.17 S	Are color codes used consistently?		
9.18 S	For workers with color vision deficiencies, are there alternative labels or codes?		
9.19 S	Are the number and combinations of colors appropriate?		
9.20 S	Are colors distinct?		
9.21 S	Is contrast used to identify and distinguish important items?		
9.22 S	Is the screen uncluttered?		
9.23 D	Can operators pan and zoom displays?		
9.24 S	Is the information logically assigned to screens?		
9.25 D	Do screens present information in a logical sequence?		
9.26 S	Is the process appropriately subdivided across screens?		
9.27 D	Are there rapid, easy, and direct navigation/linkages between pages?		
9.28 D	Is the number of pages that must be monitored manageable?		
9.29 D	Is there an appropriate balance of summary and detail screens?		
9.30 S	Are the screen designs and layout s consistent?		
9.31 S	Is screen terminology consistent?		
9.32 S	Does the screen design meet user expectations (prior experience, cultural norms, company norms, vendor norms)?		
9.33 D	Is the number of display screens sufficient?		
9.34 D	Can screens be accessed quickly and easily?		
9.35D	Are process changes easily noticed?		
9.36 D	Is it easy to confirm that the process is operating normally?		
9.37 D	Can operators diagnose faults readily?		
9.38 S	Are there sequence displays for batch and sequential processing?		
9.39 D	Is information displayed at a pace suitable for the operator?		
9.40 S	Are alarms in a suitable form and location?		
9.41 S	Are alarm indications separate from plant status indications?		

*S = static conditions; D = dynamic conditions.

Human Factors Checklist

Analysis of _____

at _____ Facility

Date Completed:_____ Completed By:_____

9.42 S, D	Are alarms prioritized?		
9.43 S, D	Are related alarms grouped?		
9.44 D	Are nuisance alarms eliminated?		
9.45 S	Are return-to-normal indications provided?		
9.46 D	Are alarm settings adjusted according to the process operating mode?		
9.47 D	Are cascade alarm suppressed?		
9.48 S	Is there a choice of alarm presentation modes (panel, VDU)?		
9.49 S, D	Are alarms suitably designed and include auto-silencing?		
9.50 D	Are there appropriate means for alarm acknowledgement		
9.51 D	Are there provisions for handling alarm flooding?		
9.52 D	Are all display/alarm sounds meaningful to the operators?		

10. Safe Havens

10.1 S	Are safe havens clearly designated?		
10.2 S	Is there sufficient room to accommodate those who may shelter there as well as any ongoing emergency response activities?		
10.3 S	Is there a designated area and supplies for those needing medical attention?		
10.4 S	Is there sufficient PPE if evacuation is required?		
10.4 S	Are escape routes designated?		
10.6 S	Are alternative shelters designated?		
10.7 S	Are shelter utilities (power, HVAC, sanitation) sufficient for the expected number of people and event duration?		
10.8 S	Are there adequate means of communication?		
10.9 S	Are workers trained in safe haven procedures?		

11. Labeling

11.1 S	Are label views unobstructed by adjacent items?		
11.2 S	Are labels on the flattest, least cluttered surfaces?		
11.3 S	Are labels on the main equipment chassis?		
11.4	Are labels positioned to minimize wear?		

*S = static conditions; D = dynamic conditions.

Human Factors Checklist

Analysis of _____

at _____ Facility

Date Completed:_____   Completed By:_____

S			
11.5 S	Are labels positioned to minimize accumulation of grease and grime?		
11.6 S	Are labels secured to prevent accidental removal or loss?		
11.7 D	Are labels easily readable from normal work locations?		
11.8 D	Are signs that warn workers of hazardous materials or conditions adequately visible and clearly understood?		
11.9 D	Are adequate signs posted in cleanup and maintenance areas to warn workers of special or unique hazards?		
11.10 S	Is all important equipment (vessels, pipes, valves, instruments, controls, etc.) clearly and unambiguously labeled?		
11.11 S	Does the labeling program include components (e.g., small valves) that are mentioned in the procedures even if they are not assigned an equipment number?		
11.12 S	Are plant instruments and controls clearly labeled?		
11.13 S	Are the labels accurate?		
11.14 S	Do the labels conform to populational stereotypes?		
11.15 S	Do the labels provide sufficient information to unambiguously identify the equipment?		
11.16 S	Do the labels correspond to the procedures and drawings?		
11.17 S	Is the responsibility for maintaining and updating the labels clearly assigned to an individual?		

12. Training

12.1 S	Are regulatory requirements and industry standards addressed?		
12.2 S	Does management send people to training as scheduled?		
12.3 S	Are both initial and ongoing training requirements established?		
12.4 S	Are job descriptions correlated with training requirements?		
12.5 S	Are performance requirements and standards established?		
12.6 S	Is the means by which trainees must demonstrate they have understood the training specified?		
12.7 S	Are standards for instructors established?		
12.8 S	Is training designed for the specific functions and jobs to be performed?		
12.9	Is site-specific training used when		

*S = static conditions; D = dynamic conditions.

Human Factors Checklist

Analysis of _____

at _____ Facility

Date Completed:_____ Completed By:_____

S	appropriate?		
12.10 S	Do operators and maintenance workers receive adequate training in safely and reliably performing their assigned tasks before they are allowed to work without direct supervision?		
12.11 S	Does a periodic refresher training program exist?		
12.12 S	Is special or refresher training provided in preparation for an infrequently performed operation?		
12.13 S	Does refresher training vary to maintain interest?		
12.14 S	When changes are made, are workers trained in the new operation, including an explanation of why the change was made and how worker safety and performance can be affected by the change?		
12.15 S	Are new knowledge, information, and lessons learned incorporated into training?		
12.16 S	Are training records maintained?		
12.17 S	Does operator and maintenance worker training include training in appropriate emergency response?		

13. Communications

13.1 S	Is there an official plant language?		
13.2 S	Is there a dictionary of standard plant terms?		
13.3 S	Is there a standard form used for communicating information between shifts?		
13.4 S	Is there a standard form used for communicating information between work groups?		
13.5 D	Is repeat-back used for oral communication?		
13.6 D	Is communications equipment adequate and easily accessible?		

14. Documentation Design and Use

14.1 S	Is the documentation media suitable for the user?		
14.2 S	Is the documentation easy to search, navigate, and use?		
14.3 S	Is the documentation available when and where needed?		
14.4 S	Is the documentation current and accurate?		

*S = static conditions; D = dynamic conditions.

Human Factors Checklist

Analysis of _____

at _____ Facility

Date Completed:_____ Completed By:_____

15. Environmental Factors

15.1 S	Is noise maintained at a tolerable level?		
15.2 S	Are vibrations damped to tolerable levels?		
15.3 S	Is the temperature normally within comfortable bounds?		
15.4 S	Is air quality maintained at a tolerable level?		
15.5 S	Is the lighting sufficient for all facility operations under normal and emergency conditions?		
15.6 S	Is the general environment conducive to efficient performance (wet, dirty, slippery)?		

16. Workloads and Staffing Levels

16.1 D	Is the physical workload sustainable over an entire shift?		
16.2 D	Are the normal work activities sufficiently demanding to maintain the operator's attention?		
16.3 D	Are workers able to perform tasks according to procedure within the time allotted?		
16.4 D	Is the staffing level appropriate for all modes of operation (normal, emergency, etc.)?		
16.5 D	Can additional staff (e.g., from other areas or from off-site) be called in quickly to help during emergencies?		
16.6 D	Are the operators only in the control room or do they do other things?		
16.7 D	Can operators perform all manual adjustments required during normal and emergency operations (not an excessive number of adjustments required)?		
16.8 D	Do the controls and indications support operator response to upset/emergency conditions?		
16.9 D	Is there sufficient time, adequate feedback, and controls for personnel to recover from errors?		
16.10 D	Are provisions in place to limit the time a worker spends in harsh environments (i.e., too hot, too cold, confined space, etc.)?		

17. Shiftwork Issues

17.1 S	Are there adequate breaks during a shift to avoid fatigue?		
17.2 S	Are there adequate breaks between shifts to avoid fatigue?		
17.3 S	Are shift rotations designed to minimize disruption of circadian rhythms?		

*S = static conditions; D = dynamic conditions.

Human Factors Checklist

Analysis of _____

at _____ Facility

Date Completed:_____ Completed By:_____

17.4 S	Have the effects of shift duration and rotation been considered in establishing workloads?		
17.5 S	Are there limits on the total number of hours worked consecutively, daily, and weekly?		
17.6 S	Are there limits on the total number of hours worked consecutively, daily, and weekly?		
17.7 S	Is the number of hours personnel work during startup, turnarounds, or high production periods limited so that worker safety and performance are not adversely affected?		
17.8 S	Is the work environment designed to minimize the impact of shift rotation?		

18. Manual Materials Handling

18.1 S	Are there diagnosed cases of musculoskeletal disorders?		
18.2 S	Do workers complain of aches or pain when/after performing certain tasks?		
18.3 S	Are there any jobs which workers avoid?		
18.4 S	Are there tasks involving large vertical movements or long carrying distances?		
18.5 S	Do people have to work in awkward postures or maintain fixed positions?		
18.6 S	Are there tasks involving repetitive motions for long periods?		
18.7 S	Are there tasks that involve strenuous pushing or pulling		
18.8 S	Are workers required to handle loads that are unwieldy, unstable, or difficult to grasp?		
18.9 S	Do workers perform heavy manual handling?		
18.10 S .	Are workers rushed to complete tasks?		
18.11 S	Are work methods and conditions within worker control?		
18.12 S	Are the PPE requirements tolerable for normal and emergency duties?		

19. Safety Culture

19.1 S	Is safety emphasized over expediency or profit?		
19.2 S	Are safety systems kept in good working order?		
19.3 S	Is management visibly involved and committed to safety??		
19.4 S	Are there frequent communications about safety and free sharing of lessons learned?		
19.5 S	Are diverse teams involved in evaluating risks?		
19.6	Are there adequate health and safety		

*S = static conditions; D = dynamic conditions.

Human Factors Checklist

Analysis of _____

at _____ Facility

Date Completed: _____ Completed By: _____

S	resources?		
19.7 S	Is there a high level of trust between management and the front-line workers?		
19.8 S	Are workers empowered to stop work if they feel unsafe?		
19.9 S	Are working areas generally clean?		
19.10 S	Do administrative controls exist to control when instruments, displays, or controls are deliberately disabled or bypassed and how they are returned to service?		
19.11 S	Are operators trained to shutdown the process when in doubt about whether it can continue to operate safely/properly?		
19.12 D	Do operators sufficiently use procedures during upset/emergency conditions?		

20. Behavior-Based Safety

20.1 S	Have critical task inventories been developed?		
20.2 S	Do workers observe and coach co-workers in safe work practices?		
20.3 S	Do workers take the initiative to correct the behavior of anyone (including contractors) who is not following safe work practices?		
20.4 S	Do workers identify the underlying causes of unsafe behaviors?		
20.5 S	Does management provide the resources to promptly correct causes of unsafe behavior?		
20.6 S	Are operators and maintenance workers trained to request assistance when they believe they need it to properly perform a task?		

21. Project Planning, Design and Execution

21.1 D	Were human factors issues identified and addressed at each phase of the project?		
21.2 D	Were the right issues identified?		
21.3 D	Were the issues identified at the correct time?		
21.4 D	Were the proper tools and processes used?		
21.5 D	Were the correct individuals engaged in the process?		
21.6 D	Was there adequate follow-up?		
21.7 D	Were the benefits of incorporating human factors in the project identified?		
21.8 D	Were the costs identified?		
21.9 D	Were steps taken to improve the human factors approaches during each phase of the project?		

*S = static conditions; D = dynamic conditions.

Human Factors Checklist

Analysis of _____

at _____ Facility

Date Completed:_____   Completed By:_____

	22. Procedures		
22.1 S	Is the procedure concise and easy to use?		
22.2 S	Is the level of detail appropriate (consider user background, job complexity)?		
22.3 S	Are conditional instructions easy to understand?		
22.4 S	If an action must meet more than two requirements, are the requirements listed?		
22.5 S	Are calculations clear and understandable?		
22.6 S	For complicated or critical calculations, is a formula or table included or referenced?		
22.7 S	Can graphs, charts, and tables be easily and accurately extracted and interpreted?		
22.8 S	Are steps written in short, concise, statements?		
22.9 S	Are the same terms used consistently for the same components or operations?		
22.10 S	Do the procedure contain plenty of white space?		
22.11 S	Do the procedures contain tab markers to help locate them quickly?		
22.12 S	Are lines or white spaces used to separate groups of related items?		
22.13 S	Is the Times Roman font used?		
22.14 S	Is the font size consistent throughout and at least 12-point?		
22.15 S	Are CAPS are used for major titles?		
22.16 S	Is mixed, upper and lower case text used throughout?		
22.17 S	Is text left justified?		
22.18 S	Are steps identified by their own unique numbers?		
22.19 S	Is each step listed in sequential order, as it should be performed?		
22.20 S	Does each step begin with an action verb?		
22.21 S	Are CAUTIONS, WARNINGS, and NOTES placed immediately before the step to which they apply?		
22.22 S	Do CAUTIONS, WARNINGS, and NOTES stand out from the procedure steps?		
22.23 S	Are P&IDs or flow charts placed ahead of the relevant steps or included with them?		
22.24 S	If conditions or criteria are used to help the user make a decision or recognize a situation, do they precede the action?		
22.25	Is the procedure written for the lowest		

*S = static conditions; D = dynamic conditions.

Human Factors Checklist

Analysis of _____

at _____ Facility

Date Completed: _____ Completed By: _____

S	education level allowed among qualified users		
22.26 S	Does the title accurately describe the nature of the activity?		
22.27 S	If the procedure is over 5 pages, is it equipped with a Table of Contents on the first page?		
22.28 S	Does the first page of the procedure state the title, objective, background, references, special equipment, and precautions, prerequisites, author, and approval?		
22.29 S	Is procedure control information included on *each* page, such as unit name or identifier, title, number, date of issue, approval date, required review date, revision number, page number, and total pages		
22.30 S	Is the last page of the procedure clearly identified?		
22.31 S	Are temporary procedures clearly identified?		
22.32 S	Does every procedure have a unique and permanent identifier?		
22.33 S	For duplicate processes, are the procedures complete and accurate for each process?		
22.34 S	Is all information necessary for performing the procedure included or referenced in the procedure?		
22.35 S	Does the procedure include all steps required to complete a task?		
22.36 S	Does the written procedure match the way the task is done in practice?		
22.37 S	Are all items referenced in the procedure listed in the "References" section of the procedure?		
22.38 S	Are items listed in the references section of the procedure correctly and completely identified?		
22.39 S	Do the references contain a list of supporting documents and locations?		
22.40 S	If more than one person is required to perform the procedure, is the person responsible for performing each step identified?		
22.41 S	Are steps that can be done simultaneously noted?		
22.42 S	Is a signoff line provided for verifying critical steps of a procedure?		
22.43 S	If the procedure requires coordination with others, does it contain a checklist, signoff, or other method for indicating the steps or actions have been completed?		
22.44 S	If a step contains more than two items, are they listed rather than buried in the text?		
22.45 S	If multiple actions are included in a single step, can the actions actually be performed simultaneously or as a single action?		
22.46 S	Are steps that must be performed in a fixed sequence identified as such?		

*S = static conditions; D = dynamic conditions.

Human Factors Checklist

Analysis of _____

at _____ Facility

Date Completed:_____ Completed By:_____

22.47 S	Are operating or maintenance limits or specifications written in quantitative terms?		
22.48 S	Does the procedure provide instructions for all reasonable contingencies?		
22.49 S	If the contingency instructions are used, does the contingency statement precede the action statement?		
22.50 S	Do procedures that specify alignment such as valve positions, pipe and spool configurations, or hose station hook-ups: specify each item, identify each item with a unique number or designator, specify the position in which the item is to be placed, and indicate where the user records the position, if applicable?		
22.51 S	Do emergency operating procedures contain provisions for verifying: conditions associated with an emergency (initiating conditions), automatic actions associated with an emergency, and performance of critical actions?		
22.52 S	Do emergency procedures provide adequate guidance on diagnosing system upsets?		
22.53 S	Can the procedure be implemented safely (without creating new risks)?		
22.54 S	Do maintenance procedures include required follow-up actions or tests and tell the user who must be notified?		
22.55 S	If someone with special qualifications must perform a procedure, are the required technical skill levels identified?		
22.56 S	Is a complete, current set of procedures available for workers to use?		
22.57 S	Do the procedures cover all operational modes in sufficient detail (startup, shutdown, operations, standby, maintenance, etc.)?		
22.58 S	Are safe operating limits documented, providing consequences of deviating from limits and actions to take when deviations occur?		
22.59 S	Are the procedures accurate (i.e., do they reflect the way in which the work is actually performed)?		
22.60 S	Is responsibility assigned for updating the procedures, distributing revisions of the procedures, and ensuring that workers are using current revisions of the procedures?		
22.61 S	Are temporary notes or instructions incorporated into revisions of written operating procedures as soon as practical?		
23.62 S	Do procedures address the personal protective equipment (PPE) required when performing routine and/or nonroutine tasks?		

*S = static conditions; D = dynamic conditions.

Human Factors Checklist

Analysis of _____

at _____ Facility

Date Completed:_____   Completed By:_____

23. Maintenance			
23.1 S	Are critical maintenance tasks identified?		
23.2 S	Is equipment designed for maintenance access?		
23.3 S	Are self-checks used during maintenance activities?		
23.4 S	Are post-maintenance checks performed to detect errors?		
23.5 S	Are the right tools available and used when needed?		
23.6 S	Are special tools required to perform any tasks safely or efficiently?		

24. Safe Work Practices and Permit-to-Work Systems			
24.1 S	Does the PTW system specify worker qualifications?		
24.2 S	Does the PTW system require pre-job briefs?		
24.3 S	Must permits be displayed in the work area?		
24.4 S	Must permits be provided to other affected work groups?		
24.5 S	Does the PTW system identify precautions and verify their completion?		
24.6 S	Does the PTW system ensure the correct equipment is worked on?		
24.7 S	Does the PTW system require independent cross checks?		
24.8 S	Does the PTW system require management authorization and supervision of specific tasks?		
24.9 S	Is the PTW system always used?		
24.10 S	Does the PTW system identify relevant hazards?		

25. Management of Change			
25.1 S	Do workers follow the MOC system, even for temporary or undocumented operational changes?		
25.2 S	Are workers trained in the MOC procedure?		
25.3 S	Are changes reviewed in a timely manner?		

26. Qualitative Hazard Analysis			
26.1 S	Were human factors issues considered in the process hazard analysis?		

*S = static conditions; D = dynamic conditions.

Human Factors Checklist

Analysis of _____

at _____ Facility

Date Completed:_____ Completed By:_____

27. Quantitative Risk Analysis

27.1 S	Has data been collected on human error rates?		
27.2 S	Has the likelihood of critical errors been analyzed?		

28. Safety Systems

28.1 S	Is there adequate margin between safety system setpoints and normal conditions?		
28.2 S	Is there a documented procedure for bypassing safety systems?		
28.3 S	Is a log of bypassed interlocks kept?		
28.4 S	Is initiation of safety systems automatic?		

29. Competence Management

29.1 S	Are competence requirements identified?		
29.2 S	Are personnel recruited and selected based on required competencies?		
29.3 S	Is worker competence initially and periodically assessed?		
29.4 S	Are competence certification records kept?		

30. Emergency Preparedness and Response

30.1 S	Is there an emergency response plan?		
30.2 S	Is there an incident command structure?		
30.3 S	Have workers been trained in their emergency response responsibilities?		
30.4 S	Is adequate emergency equipment available and accessible during the emergency?		
30.5 S	Are there reliable means for communicating in an emergency?		
30.6 S	Do operators practice emergency response under realistic conditions (i.e., while wearing emergency protective equipment, with emergency lighting, etc.)?		
30.7 S	Do operators practice emergency response during extreme environmental conditions (e.g., at night or when it is very cold)?		
30.8 S	Are periodic emergency drills conducted?		
30.9 S	Are emergency drills witnessed by observers and critiqued?		
30.10 S	Are emergency evacuation routes clearly marked?		

*S = static conditions; D = dynamic conditions.

Human Factors Checklist

Analysis of _____

at _____ Facility

Date Completed:_____ Completed By:_____

	31. Incident Investigation		
31.1 S	Are criteria for reporting incidents defined?		
31.2 S	Are incidents investigated promptly?		
31.3 S	Do trained peers investigate incidents?		
31.4 S	Do incident investigations identify root causes (management system weaknesses)?		
31.5 S	Are corrective actions implemented in a timely manner?		

*S = static conditions; D = dynamic conditions.

Index